Determinants and Implications of the Growing Scale of Livestock Farms in Four Fast-Growing Developing Countries

Christopher L. Delgado, Clare A. Narrod, and Marites M. Tiongco
  with

Geraldo Sant'Ana de Camargo Barros, Maria Angeles Catelo,
Achilles Costales, Rajesh Mehta, Viroj Naranong,
Nipon Poapongsakorn, Vijay Paul Sharma, and Sergio de Zen

  in collaboration with

Center for Advanced Studies on Applied Economics
Food and Agriculture Organization of the United Nations
Indian Institute of Management
Livestock, Environment, and Development Initiative
Research and Information System for Developing Countries
Thailand Development Research Institute
University of the Philippines Los Baños

RESEARCH REPORT 157

Copyright © 2008 International Food Policy Research Institute. All rights reserved.
Sections of this material may be reproduced for personal and not-for-profit use
without the express written permission of but with acknowledgment to IFPRI. To
reproduce material contained herein for profit or commercial use requires express
written permission. To obtain permission, contact the Communications Division
<ifpri-copyright@cgiar.org>.

International Food Policy Research Institute
2033 K Street, NW
Washington, D.C. 20006-1002, U.S.A.
Telephone +1-202-862-5600
www.ifpri.org

DOI: 10.2499/9780896291669RR157

Library of Congress Cataloging-in-Publication Data

Delgado, Christopher L.
  Determinants and implications of the growing scale of livestock farms
in four fast-growing developing countries / Christopher L. Delgado,
Clare A. Narrod, Marites M. Tiongco ; with Geraldo Sant'Ana de Camargo
Barros . . . [et al.].
     p.   cm.—(IFPRI research report ; 157)
  Includes bibliographical references.
  ISBN 978-0-89629-166-9 (alk. paper)
  1. Livestock systems—Developing countries.   2. Livestock
productivity—Economic aspects—Developing countries.   I. Narrod,
Clare A.   II. Tiongco, Marites M.   III. Barros, Geraldo Sant'Ana
de Camargo.   IV. International Food Policy Research Institute.   V. Title.
VI. Series: Research report (International Food Policy Research Institute) ; 157.
SF55.D44D45   2008
636.009172'4—dc22   2008004961

Contents

Tables

Figures

Foreword

Human population growth, increasing urbanization, and rising incomes have fueled unprecedented growth in the demand for livestock products in the developing world. This creates pressures on world feedgrain and oilseed markets, but presents opportunities for poor people who rear, process, or market livestock products to enhance their livelihoods. However, large-scale, industrialized forms of poultry, pig, and dairy production are also growing rapidly to compete for these new markets, and poor people may find it increasingly difficult to compete. Additionally, the growing concentration of animals in large units near cities is associated with greater pollution and increased risk of transmission of diseases, including to humans. Finally, the transformation of traditional animal husbandry into an industrial process raises ethical issues in places where adequate safeguards for animal welfare are not fully developed.

To better understand why the average size of livestock operations in developing countries is growing and what the implications of this are for smallholder incomes and the environment, the International Food Policy Research Institute has partnered with the Food and Agriculture Organization of the United Nations (FAO) to study the competitiveness of smallholder livestock production systems vis-à-vis large-scale systems operating in the same environment. Using farm-level data, the study explores the relationships among average farm sizes, the profit efficiency of individual enterprises, economies of scale in input supply, policy distortions favoring larger enterprises, negative environmental externalities, and transaction costs.

Results show that smallholder livestock producers can still compete with larger producers in many of the cases studied under current market conditions, despite frequently paying higher prices for inputs due to economies of scale, receiving lower prices for their output due to higher transaction costs, and internalizing more of their environmental costs. This is because of the savings of smaller units on overhead items, lower labor costs per unit, and possibly more intensive supervision, leading to relatively high profit efficiencies. Strengthening producers' associations and promoting contract farming may help smallholders deal with events beyond the farm gate—such as growing retailer demand for consistency and demonstrable safety—that might eventually displace them even if they are relatively efficient producers.

Public policy can play a role in assisting private-sector firms to reduce the transaction costs that increasingly exclude smallholders from participating in growing livestock markets.

Joachim von Braun
Director General, IFPRI

Acknowledgments

The present report synthesizes results of a joint IFPRI–United Nations Food and Agricultural Organization (FAO) research project carried out from 2001 to 2003 titled Livestock Industrialization, Trade and Social-Health-Environment Impacts in Developing Countries. The report was subsequently refined. From September 2003 to April 2006, Christopher Delgado and Marites Tiongco were part of the International Livestock Research Institute (ILRI) / IFPRI Joint Program on Livestock Market Opportunities, and the authors are grateful for ILRI's support of their research during this time. The collegial interaction of colleagues at the International Livestock Research Institute, including co-authorship of the Indian dairy case study, is gratefully acknowledged. Research was conducted via collaborative case studies led by national institutions in Brazil, India, the Philippines, and Thailand. Intellectual ownership of insights specific to the countries covered by the issues papers belongs to the authors of the country reports,[1] who nevertheless should not be held responsible for any errors of interpretation on our part. The strong collaboration of Pierre Gerber of FAO and the Livestock, Environment, and Development Initiative (LEAD) in providing the many complex maps of livestock distribution used by the project is very much appreciated.

The authors also acknowledge the valuable insights and suggestions received from Henning Steinfeld of FAO/LEAD and Cees de Haan, then of the World Bank. They also thank Nikolas Wada and Gloria Paniagua, then of IFPRI, along with a series of external editors organized by Uday Mohan of IFPRI, for valuable editorial assistance. They further thank Christine Ellefson and Laura Contaldi of FAO's Livestock Information Sector Analysis and Policy Division for administrative support and logistical arrangements for the research and Tigist Defabachew, Shirley Raymundo, and Joy Fabela of IFPRI for administrative support for the research and for arranging the many research visits of collaborators to Washington. All concerned will fondly remember Shirley Raymundo and Joy Fabela for their efforts beyond the call of duty in organizing and formatting more than 700 pages of the country reports that the present manuscript draws on, in addition to Shirley's work on the present manuscript. They are also very appreciative of the help of Reno Dewina, who facilitated the present report's empirical analysis by assisting in the data extraction for the various country studies.

Finally, the authors gratefully acknowledge very detailed and perceptive reviews by two anonymous internal and two anonymous external peer reviewers, all of whom have provided the team with significant extra work that has significantly improved the report and, they hope, offered commensurate extra benefit to the final reader.

---

[1]The leaders of the various country papers we draw on are listed as co-authors on the cover of this report.

Summary

The research underlying this report investigated factors affecting the scaling-up of selected livestock products in Brazil, India, the Philippines, and Thailand, with particular attention directed at understanding issues affecting small-scale producers.[2] *Scaling-up* is defined here as growth in the average size of farm in terms of annual livestock sales. The urgency of the topic is driven by the fact that livestock are among the few commodities that smallholder farmers widely produce that are growing rapidly in demand, and thus the interest for poverty alleviation is strong. However, there are signs that smallholders may ultimately be displaced from this source of livelihood by competition from larger-scale farms. Furthermore, the rapid growth in production of pigs and poultry has also been associated with significant environmental problems in the zones of most rapid growth. The interactions between overall growth in production, size of farms, and sustainability are crucial to rural poverty alleviation in a sustainable manner.

Therefore, the objectives of the research were to assess whether the market share of large farms is growing relative to small farms and, if so, why. The analysis is directed primarily toward the issues of efficiency in production and its determinants, but also deals with the unit costs of inputs received and the unit prices for outputs obtained. The study specifically does not investigate the likely existence of economies of scale in output marketing, which are anecdotally thought to be large and will over time largely work to the disadvantage of independent (nonintegrated) farmers, especially small-scale ones. This important topic is deferred to a different study on contract farming in poultry and dairy using different data and approaches.

The research was pursued by setting up models and collecting farm-level data to test the following hypotheses:

- Small-scale producers have profits per unit of output that are higher than or equal to those of large producers.
- Small-scale producers are more efficient users of farm resources to secure profits than are large producers, other things equal.
- Small farmers expend a greater amount of effort or investment in abatement of negative environmental externalities per unit of output than do large farmers.
- Uncompensated negative environmental externalities favor the nominal profit efficiency of large farms over small ones, ceteris paribus.
- Profits of small producers are more sensitive to "transaction costs" than are those of large-scale producers.
- Contract farmers are more profit-efficient than independent farmers at comparable scales of operation.

---

[2]The commodity samples are: Brazil—broilers, eggs, swine, and milk; India—broilers, eggs, and milk; the Philippines —broilers and swine; and Thailand—broilers, swine, eggs, and milk.

Chapter 2 looks at evidence of growth and concentration of the livestock sector and establishes that scaling-up in livestock production is in fact occurring in all four countries studied. It suggests that production in the 1980s and 1990s tended to concentrate around capital cities and other major demand centers with the exception of Brazil. Since the late 1990s, larger-scale farms have been increasingly locating in remoter areas farther away from population centers, disease, and environmental challenges. Smallholder output continues to grow at high rates in certain cases, such as that of dairy farmers in India and swine producers in the Philippines, but the output from large-scale enterprises is growing even more rapidly, taking market share away from smallholders. This is particularly relevant in Brazil and Thailand, where there are relatively small numbers of small-scale producers left in the broiler business and the role of smallholders in both dairy farming and swine production is shrinking in relative terms. The question, then, is whether large-scale livestock production will outcompete that of smallholder producers everywhere and eventually provoke their exit from the sector.

To analyze the scaling-up of livestock production, Chapter 3 defines a quantitative measure of relative farm competitiveness in production based on the use of stochastic profit frontiers estimated for each country and commodity, with technical inefficiency effects included, following the approach of Battese and Coelli (1995). It looks at factors that differ across farms that might explain higher relative competitiveness of specific farms. These factors include the usual determinants of profit efficiency, such as prices faced by the farm and fixed farm resources not traded in markets, asymmetries in access to assets (credit, liquidity, fixed capital) and information (education, experience, communication facilities), externalities (some farmers get away with uncompensated pollution, while others do not), and policies (some get a better deal from the state than do others). Chapter 4 describes field surveys that were conducted on cross-sections of farms of different sizes (small-scale, medium-scale, and large-scale or commercial) and type of production arrangement (independent and contract).

Chapter 5 summarizes the descriptive results of the survey and seeks to establish through descriptive statistics scale-related differences in farm-specific transaction costs, such as access to information and input and output markets, to direct further analytical work on explaining farm profit per unit of output. Results show that small-scale producers across the countries studied are characterized by small landholdings, low levels of education, few years of experience in livestock production, moderate use of inputs, and limited access to input and output markets. Large-scale farm households, therefore, may have a better chance of overcoming high transaction costs than do small-scale producers. On the other hand, smallholders have a chance to compete with larger-scale producers if they can cost family labor at less than the full opportunity cost of hired labor doing the same tasks on larger farms, perhaps because much of the labor on small farms consists of extra tasks performed by women or children who are at home for other reasons. The study does not claim that the opportunity cost of family labor is zero, but it repeats the relevant calculations with family labor costed either at full market wage rates or not at all, with the correct answer postulated to fall somewhere in the resulting range between the two.

Chapter 6 first discusses the likely impact of scaling-up of livestock production on the environment, then investigates the issue of whether large-scale sample farms make less of an effort per unit of output to mitigate pollution through expenditures on lagoons, manure spreading on farm land, and other mitigation activities than do small-scale sample farms in the study countries. If so, by hypothesis they would be polluting more per unit of output than are small farms, assuming that each unit of output produces the same effluent regardless of production system. Results show that across countries and commodities, small-scale farms have higher environmental mitigation "expenditures" per unit of output than large-scale farms. The exceptions are poultry farmers in Thailand, where large-scale operations in the densely settled export-certified zone spend more per unit than smaller farms, and dairy farmers in Thailand, where the larger-scale farmers have more crop land per animal than do the smaller-scale farmers in the sample. Results are backed up by a second, more conventional procedure that estimates mass balances of nutrients per hectare added to farm soils on different farms. Again, it is shown that large farms load more excess nutrients per hectare of land than do small farms, suggesting that in fact large farms benefit from a higher uncompensated environmental externality per unit of output than do small farms, giving them a distorted cost advantage.

Chapter 7 reports the econometric evidence from stochastic profit frontier analysis using the Battese-Coelli (1995) approach to technical inefficiencies on why some farms are relatively more profit-efficient than others at making farm profits from livestock. Chapter 8 then ties these results together with the preceding descriptive analysis. The main findings are as follows:

- Independent small farms in India and the Philippines typically have higher profits per unit than do independent large farms, but this is not the case for large and small contract farms. On the other hand, large independent farms are relatively more profit-efficient than small independents almost everywhere, suggesting that over time they will continue to outcompete smallholders and gain further market share. India, where most farms are small and dairy dominates, is a notable exception.

- In Brazil and Thailand in particular, where scaling-up has already occurred to a large extent in monogastrics,[3] the outlook for independent smallholder producers of the items studied is not rosy. In the Philippines, there is considerable scope on economic grounds for the improvement of smallholder farming through vertical coordination of those farms that are presently a little larger than the bottom 20 percent of the size distribution of farms, especially for swine producers.

- Small farms have less of a negative impact on the environment than do large farms. Hence, environmental concerns are compatible with promoting small-scale livestock production. Large farms that are more environmentally responsible are also more competitive within the class of large farms (particularly Philippine broiler and swine farms). Efficient large-scale production is compatible with good environmental behavior, although it seems likely that both are correlated with some third factor, such as the chicken and swine inventory held on-farm or the ability to sell manure produced on-farm. It seems plausible that over time enforcement of environmental regulations will be more similar to enforcement of health regulations, and all producers will be forced to bear the same costs regardless of size. While a level environmental playing field will probably benefit small farms more than large ones, it will not by itself reverse the profit-efficiency advantage of large farms over small ones.

- Beyond maximizing the use of (assumed) lower-opportunity-cost family labor, the relative competitiveness of smallholders is largely determined by farm-specific abilities to overcome barriers to information and assets, such as credit and market information. Therefore, a possible key to pro-poor livestock development is institutional development that overcomes the disproportionately high transaction costs that smallholders face in securing quality inputs and getting market recognition for quality outputs. This is also in line with the finding that as markets gravitate to higher-end concerns of quality, safety, and reliability of supply, smallholders will have to be associated with institutions that can supply the technology, inputs, information, and accreditation necessary for competing in higher-value markets.

---

[3]*Monogastrics* are animals with one stomach compartment; examples are pigs and poultry.

- Contract farmers have higher profits per unit of output than do independent farmers in some but not all cases, and they tend to be more profit-efficient than independent farmers at all scales (except for small swine farms in the Philippines). Therefore, contract farming has real potential to better help incorporate smallholders in high-value supply chains that require specialized inputs and sell to markets for specialized outputs. However, the country studies reveal that contract farming covers a multitude of arrangements, some of which are more beneficial to smallholders than others.
- Contract farming is shown to be even more useful in improving the competitiveness of large farms than in improving that of small ones. From the integrator standpoint, there are also likely to be cost savings in dealing with a few large farms than in dealing with lots of small ones. Therefore, the policy environment for contracting will be especially important to its usefulness as a tool for poverty alleviation.

Overall, the study concludes that in many cases smallholders other than the smallest backyard producers will be able to stay in the livestock business for a long time. However, if the opportunity cost of family labor rises and begins to approach local market wage rates, as is beginning to happen in periurban areas of China, for example, much of the competitiveness of smallholder operations vis-à-vis large farms will be vitiated. Furthermore, emerging disease threats and environmental backlash suggest that the future of large and small producers will sink or rise together in a context of growing demand and the ability to act collectively to deal with emerging threats. Finding ways to increase synergies between the two groups is very much in the social interest. Finally, there is reason to believe that there are significant economies of scale in livestock product distribution as supply chains become longer, wider, and more anonymous, although this is not investigated in the present study. If so, the future for independent livestock farmers, whether large or small, will eventually depend on the options for integration with input supply and output marketing operations, which must necessarily be the subject of a different study.

CHAPTER 1

Introduction

The world is entering a period of rapid change in the way that animal products are produced, processed, marketed, and consumed. At the farm level, the trends that increasingly have been observed in developed countries—scaling-up of production and increased concentration of larger-scale operations with increased environmental problems (Abdalla, Lanyon, and Hallberg 1995)—are becoming apparent in the developing countries (Sere and Steinfeld 1996; Steinfeld, de Haan, and Blackburn 1997; Nell 1998). The number of large-scale industrial livestock enterprises has increased steadily since the 1970s in developed countries, while the rate of growth of large-scale production has outstripped the growth of small-scale producers (Martinez 2002; Morrison Paul et al. 2004).

It is plausible that the same trend will occur in developing countries, especially for poultry and swine. The issues of concern in developing countries arise out of the much higher share of the population who are small-scale livestock producers, their lack of attractive employment options, and the sheer speed with which change is occurring. Livestock production in developing countries as a whole, and most particularly in Asia and Latin America, has been steadily transiting over the last 25 years from a multipurpose activity producing food, savings, traction power, hides, and manure with family labor and farm-produced feed on smallholder crop farms to a more specialized enterprise using hired labor, borrowed capital, and purchased inputs in systems producing more uniform-quality food items under industrial modes of organization more reminiscent of factories than of traditional smallholder farms (Sere and Steinfeld 1996; Delgado et al. 1999). This transformation is overwhelmingly associated with the growth of poultry and pig production and is referred to in this research report as "livestock industrialization." The growing scale of poultry and swine farms in developing countries will be taken in this research report to mean the movement toward livestock industrialization, and for simplicity is taken as synonymous with it.

Main Objective, Assumptions, and Approach

This study investigates factors affecting scaling-up in the sense of growth in mean number of animals per farm kept at any one point—of livestock farms in Brazil, India, the Philippines, and Thailand, four fast-changing countries with growing metropolitan areas, rapidly rising consumption of animal-source foods, and significant livestock production coming from a diverse mix of farm sizes. The main purpose of the study is to examine whether smallholder livestock producers in these countries—and, by extension, in other developing countries—can hope to stay in business over the medium to long run when faced with competition from large-scale and increasingly industrialized livestock operations.

Two untested assumptions underlie the analysis: (1) in market-oriented systems, non-competitive entities will eventually move to some other pursuit or at least lose significant market share, and (2) factors that promote increasingly large sizes of farm also progressively promote livestock industrialization as defined earlier. With regard to the first assumption, structural factors such as an entrenched peasantry or institutional factors such as property rights systems could slow down changes in size relative to optimal farm sizes dictated by profitability (Gardner and Pope 1978; Koester 2004). Yet it seems reasonable to assume that over time units that are not competitive will not stay engaged in the sector, or at least will not expand at a time when the competition is growing rapidly. With regard to the second assumption, it seems likely that changing consumer and wholesale demand toward safer, more reliable, and more predictable products (in terms of attributes) could speed up the process of industrialization per se, as opposed to simply larger farms, but that must necessarily be the topic of another, larger, study. The present work is focused solely on whether small farms can compete with large, as opposed to the separate and difficult question of whether they can produce items that meet the industrial demand for confidence in safety and reliability.

The approach is to establish a measurable yardstick of relative farm competitiveness in livestock, construct a view of what drives relative farm competitiveness, assess the competitiveness performance of farms of different sizes from samples collected in the four countries, and then empirically decompose the separate effects of the chosen drivers on relative competitiveness. *Competitiveness* is defined here as the ability to stay in business over time, under the assumption that in market-oriented systems those who can produce items more cheaply at a given common level of input and output prices will eventually prevail over more costly suppliers if there are no barriers to market entry or exit.

A first, necessary, condition for a farm to be competitive over time is that it is a relatively efficient producer of profits compared to other farms producing the same product in the same area, and faced with a common set of input and output prices and a common set of fixed resources. Real profit efficiency in this sense includes both technical efficiency and allocative efficiency in production (Coelli, Rao, and Battese 1998; Kumbhakar and Lovell 2000). Inefficient farms in this sense in market-oriented systems can and will (by assumption) eventually be undercut to the point where they can no longer make a profit.

In the context of developing countries, provided that the small farms are relatively efficient, there is a second, sufficient, condition for them to be competitive vis-à-vis larger farms. This is that the small farms have higher or equal unit profits than large farms in equilibrium. This is because, in the context of most developing countries, farms have such small output and are sufficiently constrained from borrowing that farmers need to cover a minimum threshold profit per unit of output to stay alive, whereas large farms need only to cover average total cost, and can if necessary even stay in business for some time covering only average variable cost, something not possible for people who need to feed their families from farm profits. In other words, even a less efficient large farm sector could drive smallholders out of the livestock business if they can undercut the minimum necessary profit per unit of smallholders for survival by operating at close to nonprofit levels. This behavior is not likely to be feasible.

In sum, the yardstick of competitiveness for smallholders in livestock production applied in this study is that both of the following conditions obtain: (1) they are more efficient at generating profits from a fixed kit of resources and facing a given set of prices than are large-scale farms and (2) in equilibrium their average unit profits are higher than those of large farms such that the latter do not have much latitude to drive them out by driving their own profit margins to zero. Other factors, such as economies of scale in procurement and processing (Ollinger, MacDonald, and Madison 2005), changing consumer preferences for a more uniform industrial product (Martinez 2002), and supply-driven concentration of procurement through investments in supermarket consolidation (Reardon and Timmer 2005) could also affect the viability of smallholder farming. The present study is necessarily limited to assessing whether production factors alone, having primarily to do with events at the farm level (technology, management, access to resources and information, and so on) will doom smallholders in livestock over time. The other topics are separate large studies on their own.

Possible Drivers of Livestock Industrialization

Technological Change

It has been suggested that the principal reason for the exit of smallholders from livestock production in developed countries is that they are not competitive with the larger operations that benefit from both technical and allocative economies of scale embodied in genetic improvement of animals and feeds or improved organization, especially for poultry and pigs, where profitable adoption simply requires larger farm sizes (Gardner and Pope 1978; Narrod 1997; Martinez 2002; Morrison Paul et al. 2004). This is a particularly difficult issue for smallholders, because it conveys a sense of inevitable economic doom over time, propelled by irreversible technological progress. Anecdotal experience suggests that many livestock production experts do not go much beyond this explanation in assuming the inevitability of livestock industrialization in developing countries.

Economies of Scale in Input Supply

The importance of understanding cost advantages in input procurement is heightened by the growing importance of monogastric livestock in total meat production costs in developing countries and by the fact that up to 70 percent of the cost of production of monogastric livestock is feed costs (Delgado et al. 1999). If larger-scale farms can regularly secure access to feed of a given quality at a lower price per unit, they gain a great cost advantage over small-scale producers. This could arise simply because it costs less per kilogram to deliver a full truckload of feed to a farm than to deliver one bag. Yet while there are undoubtedly true economies of scale to be reaped in bulk purchasing of inputs, these need to be distinguished from other feed-related cost advantages of large farms.

If large, vertically integrated operators avoid paying sales taxes on feed but small independent producers do not, small independent production is at a disadvantage because of a policy distortion (discussed later) and not an economy of scale. Similarly, feed concentrates by definition have multiple ingredients, many of which cannot be ascertained by looking at the final product. Where public enforcement of the truthfulness of ingredient labels is lax or branding is unreliable, as is often the case in developing countries, producers who are large enough to mix their own feed are more assured of receiving good-quality inputs. Put differently, smallholders face a higher "transaction cost" in terms of having less information about what their feeds contain or, put differently, they would need to spend more per unit to know in advance what their animals are actually eating. This "transaction cost" is also not a true economy of scale but a form of structural barrier specific to individual farms (discussed later).

Policy Distortions and Externalities

As we hinted earlier in the case of differential input taxation of large and small farms, another driver of growing farm sizes might be policy-induced distortions (Gardner and Pope 1978). Examples of policy distortions include both scale-specific subsidies and scale-specific differences in uncompensated negative environmental externalities. An example of the former would be if policy deliberately targeted direct subsidies such as those on credit to larger farms within a strategy of favoring rural notables for political reasons (Bates 1981).

A different sort of policy distortion affecting the incentive to scaling-up would be the case if large farms "get away with" more pollution downstream per unit of output than do small farms. If farmers reap the benefits of production but do not compensate neighbors for negative environmental externalities such as bad odors, flies, and polluted water, they are gaining a cost advantage relative to more eco-friendly or considerate farmers who incur less negative externalities. An incentive for scaling-up would exist if the net uncompensated negative externalities from livestock production benefit large producers more per unit of output than they do small producers. An example would be the dumping of large amounts of waste into watercourses by larger farms unable to absorb more manure on their fields, presumably unlike smallholders, whose ratio of stock to cropped land on-farm is typically much lower (Steinfeld, de Haan, and Blackburn 1997).

Transaction Costs Arising from Asymmetries in Information

As hinted earlier in the case of the hard-to-know composition of concentrate feeds, other important factors beside technology-driven economies of scale and policy distortions may provide incentives to scaling-up in the sense of individual farms' becoming larger. An important group of these are manifested as "transaction cost barriers" to smallholder participation in markets, where transaction costs are the hidden extra costs of search, bargaining, monitoring, and enforcement of exchange.

Transaction costs in their purest sense are the costs of exchange that arise from asymmetries across market actors in access to information (Williamson 1989). Transaction costs arise if buyers and sellers cannot know the same important pieces of information about what is being sold at the time of sale and what the alternatives are. On the output market side, if both buyers and sellers can easily ascertain the quality of an item being sold at the time of sale and prices in alternative markets, competitive forces would eventually equal market prices across different categories of farmers. But if buyers cannot be sure of the true quality of the good they are purchasing, for example, they will presumably be less willing to pay a premium for it based on claimed quality.

Smallholders that have trouble selling milk outside the local market provide an example of the latter situation, because purchasers in anonymous markets cannot be sure without a bacteriological test that unbranded milk is safe. Large-scale producers and cooperatives of small-scale ones, however, may be able to establish trust and reputation in markets, because they will be able to depend on repeat sales of quasi-branded product to the same clients, who can identify the source of the milk (Staal, Delgado, and Nicholson 1997). The clients can judge the quality of the next purchase based on a history of purchases from the same farm or co-op. Similarly, the use of poor-quality feeds for hogs produces off flavors in the meat that can only be detected when eating the final product, and poor genetics can lead to excessive back fat that is not observable until the animal is slaughtered (Payne and Wilson 1999). For this reason, larger producers may also be able to get higher prices per unit through regular large sales by developing a steady clientele that gains confidence in the quality of the product (Staal, Delgado, and Nicholson 1997). Generally, rising demand for food safety and quality is likely, ceteris paribus, to exacerbate transaction costs in livestock product exchange arising from asymmetries in access to information.

Transaction costs arising from asymmetries in information are especially prevalent in the livestock product business and represent net social losses in the sense that neither buyers nor sellers gain from their presence (Williamson 1989). They are real costs incurred as lower prices received or higher prices paid. They are likely to play a growing role in the displacement of smallholders as markets become more demanding in terms of information about the quality of the product at the time of sale (Holloway et al. 2000).

Differences in market power between large and small farms can also be thought of as transaction cost differences between large and small farms arising from differences in access to assets (De Janvry, Fafchamps, and Sadoulet 1991). A feed seller may discount feed prices to larger farms if so doing ensures continued business with an important customer. A key element of this form of transaction cost facing small farms is that they are farm-specific and not product-specific (De Janvry, Fafchamps, and Sadoulet 1991; Key, Sadoulet, and de Janvry 2000). Whether the underlying issue is assets of information is often moot. Large-scale producers are typically more creditworthy because their ability to repay is more easily known. The capital-intensive nature of livestock farms that must regularly buy stock and feed well in advance of selling the final product suggests that asymmetry in access to assets, even if it arises from asymmetries in information, is likely to make asset-poor livestock farms more subject to unfavourable deals on both the input and the output sides.

In effect, transaction costs arise from the existence of market failures for certain households; the examples mentioned earlier are failures of markets for information or credit. When there is no exchange of information and credit on these farms due to market failure, the shadow price of information or capital (as the case may be) is in effect the relevant transaction cost, but it is by definition not directly observable (De Janvry, Fafchamps, and Sadoulet 1991). Even when manifested as price differentials in different transactions, the differentials could be explained by a variety of other factors, such as quality differences or even differences in the daily aggregate supply and demand presented on the market at differ-

ent moments. A typical approach, then, is to proxy the underlying asymmetries in information and assets across households by the use of household-level variables such as education, production experience, or wealth. These household differences in proxy variables are then used in conjunction with non-transaction costs related to explainers of price outcomes (such as time of day or distance to market or observable quality differences) to explain why some transactions occur at more remunerative prices to a farmer A than to farmer B (Staal, Delgado, and Nicholson 1997).

The market failures associated with transaction costs and affecting smallholders in developing countries are usually thought of as leading to the nonindependence of household decisions concerning production and consumption (Singh, Squire, and Strauss 1986). This framework has the interesting corollary for our present purposes that rational smallholders can be shown to supply labor to their own farms for returns well below what would be required to get them to work on someone else's farm, even assuming rational utility maximization (Lopez 1986; Rosenzweig 1988; De Janvry, Fafchamps, and Sadoulet 1991).

Empirical as opposed to theoretical substantiation of the willingness of farm households to work on their own farms for marginal revenue products lower than the local market wage is inconclusive, with different studies coming to different results (Rosenzweig 1988). Feder (1985) provides support using Indian data. Tokle and Huffman (1991) provide support for the willingness of U.S. Midwest farmers to work for less at home. Abdulai and Delgado (1999) demonstrate in the case of Ghana that the supply of peasant labor to off-farm wage work is highly influenced by the job profile of one's spouse and that women were less likely to work off-farm, ceteris paribus. Because smallholder livestock-keeping is overwhelmingly carried out by women of child-raising age in developing countries, it is quite plausible that such farmers cost their own labor time at home caring for a few animals in the backyard at significantly less than market wages for casual labor off-farm.

This view for livestock is very much in keeping with the classical literature over three decades surveyed by Mellor and Mudahar (1992) explaining the widely observed inverse relationship between farm size and crop productivity in Asia. A variety of surveyed authors in different Asian countries explained the higher relative productivity of smaller farms by the increased family labor input per hectare on small farms, presumably at a lower shadow wage than the market wage rate. This finding is consistent with the findings of nonhomogeneity of family and hired labor in Asian agriculture for the purposes of productivity analysis (Deolalikar and Vijverberg 1987). It is also consistent with the work for Pakistan of Ali and Flinn (1989), who apply the concept of profit efficiency, combining both technical and allocative efficiency, to Pakistani smallholders.

There is, on the other hand, little justification for the view that family labor has zero opportunity cost. In most cases people have alternative ways to produce income, and in any event it seems likely that, other things equal, many people would prefer leisure to work activities. The approach to be adopted in the present work is that unpaid family labor input to livestock production at home has a shadow price less than the market wage rate for hired labor doing similar tasks, but is above zero. Most results will be computed using both an assumption of zero opportunity cost of family labor and again costing family labor at a market wage rate, to produce a range of results, with the right answer lying somewhere inside the range.

Family labor input per unit of output is typically highest on the smallest farms. Thus, if such farms compare unfavourably in performance to larger farms while still not costing family labor, it is not necessary to inquire much further, because even assumptions that were likely to give the small-

est farms the best showing relative to other farms were still not enough to show them as competitive. In other words, assuming zero opportunity cost for family labor only strengthens conclusions that are unfavorable to the relative efficiency of smallholders. Conversely, a finding that smallholders are more competitive than large farms when family labor is not costed is suspect, for the assumption tends to favor the estimated performance of small farms more than that of big ones.

Cost Savings from Vertical Coordination

Vertical coordination arises when different elements of the animal supply chain are integrated through some form of contractual arrangement. Cost savings, for example, could occur from purchasing inputs or selling outputs in bulk (economies of scale), from avoiding taxation of inputs (policy subsidy), or from reduction of the transaction costs between actors along the supply chain. Martinez (2002) suggests that some of the reasons for vertical coordination in the pork and poultry sectors in the United States were as follows: "Contracting and vertical integration produced a means for reducing transaction costs associated with relationship-specific transactions, especially in regions of expanding production. Contracts would provide some safeguards against opportunistic behavior, and vertical integration eliminated the exchange relationship. For attributes that are difficult to measure, gaining additional control over related production inputs may reduce measuring costs by reducing the need to measure quality" (Martinez 2002, iii).

It is thus reasonable to suppose that the costs of administering contract farming, say, with a few large farmers might be lower per unit of output than administering the same scheme with a large number of small farms. This has been offered as the principal explanation for the very rapid rise in hog farm sizes in the United States in the 1990s, when the share of independent farms in U.S. hog output fell from three quarters to one quarter of the farms in less than a decade at the same time that rapid increases in hog farm size occurred (Martinez 2002). In this case, the benefits of vertical coordination are themselves a driver of increasing farm size. However, in most cases in developing countries it would be more appropriate to think of the underlying transactions costs facing individual farms, especially smaller ones, as the driving forces that provide the incentive for vertical coordination at the same time as they increase the competitiveness of larger operations relative to smaller ones.

Why Scaling-up of Production Matters for Pro-Poor and Sustainable Growth Policies

Livestock Is the Main Game Outside Town

From the beginning of the 1970s to the mid-1990s, consumption of meat in developing countries increased by 70 million metric tons (MMT), almost triple the increase in developed countries, and consumption of milk by 105 MMT of liquid milk equivalents, more than twice the increase that occurred in developed countries. The market value of that increase in meat and milk consumption totaled approximately US$155 billion (1990 US$), more than twice the market value of increased cereal consumption under the better-known supply-led "Green Revolution" in wheat, rice, and maize. The population growth, urbanization, and income growth that fueled the increase in meat and milk consumption are expected to continue well into the new millennium, creating a veritable "Livestock Revolution" (Delgado et al. 1999).

The Livestock Revolution in meat has been most evident in East and Southeast Asia, particularly China and Brazil, as shown in Table 1.1. The share of the world's meat consumed in developing countries rose from 36 to 67 percent. Pork and poultry

accounted for 60 percent of the large net consumption increase of meat in developing countries from 1990 to 2005. The big actors in the milk story are India and Brazil. The share of developing countries in producing the world's milk rose from 34 to 47 percent (Table 1.1).

Opportunities and Threats for Smallholders

A critical issue raised by these trends is that for once a sector on which the poor are heavily involved is growing. Table 1.2 shows that in fact pork and poultry are the prominent growth sectors of developing countries' agriculture. If the poor fail to participate, they will have missed a tremendous opportunity to improve per capita livelihoods. If they participate, farm income could rise dramatically, but the conditions under which this could occur are still undetermined. Furthermore, whether the seemingly unstoppable growth of livestock demand is a good or bad thing for the poor will also depend on the environmental and public health impact of rapidly rising livestock production in close proximity to where the poor live (Delgado et al. 1999).

Some reasons that small farmers could have a cost advantage in producing high-value meat, egg, and milk commodities relative to large-scale producers are as follows (Delgado and Minot 2003):

- If farmers are more willing to work at home or have lower transaction costs in working at home on their own farm activities than for pay elsewhere, it suggests that the implicit wage rate for family labor is generally below the prevailing wage rate for agricultural labor, particularly for tasks requiring only momentary but frequent attention throughout the day.
- The family labor used by small farmers is likely to be more motivated and requires less monitoring than hired labor used by large-scale farms, so small farms are better able than large ones to

Table 1.1 Consumption of meat and milk by region, 1982–84 and 2003–05

	Total meat consumption					Total milk consumption				
	1982–84		2003–05		2020 (projected)	1982–84		2003–05		2020 (projected)
Region	Million MT	Percent of world consumption	Million MT	Percent of world consumption	Percent of world consumption	Million MT	Percent of world consumption	Million MT	Percent of world consumption	Percent of world consumption
China	16	12	71	33	32	3	1	25	5	4
India	3	2	5	3	3	34	10	73	14	20
Other East Asia	1	1	2	1	0	1	0	2	0	0
Other South Asia	1	1	3	1	2	11	3	26	5	6
Southeast Asia	4	3	11	5	6	4	1	7	1	2
Latin America	15	11	29	14	14	35	10	57	11	13
Brazil			13	6				21	4	
West Asia and North Africa	5	4	8	4	4	21	6	31	6	6
Sub-Saharan Africa	4	3	6	3	3	12	3	18	4	5
Developing world	50	36	133	67	65	122	34	240	47	57
Developed world	88	64	78	33	35	233	66	272	53	43
World	**138**		**212**			**355**		**513**		

Sources: Calculated from data in the FAO Statistics Database, http://FAOSTAT.FAO.ORG/default.htm; the 2020 projections are taken from Delgado (2003), based on historical FAO data and updated projections from the July 2002 version of IFPRI's International Model for Policy Analysis of Agricultural Commodities and Trade (IMPACT) model.

Notes: Data are three-year averages centered on the years shown. "Consumption" is direct use as food, uncooked weight, bone-in. "Meat" includes beef, pork, mutton and goat, and poultry. "Milk" is milk and milk products in liquid milk equivalents. Metric tons (MT) and kilograms are 3-year moving averages centered on the years shown.

Table 1.2 Production growth rates (in volume) in developing countries, 1975–2005

Product	Growth in volume (percent per year)
Cereals	2.2
Fruit	3.9
Vegetables	5.1
Fish	1.6
Milk	4.0
Pork	6.0
Poultry	7.0

Source: Calculated from data in the FAO Statistics Database, http://FAOSTAT.FAO.ORG/default.htm. Notes: "Fish" includes marine and freshwater fishes; "poultry" includes chicken, duck, and turkey meat.

apply careful husbandry and respond to problems in the field.

- Small livestock producers may face lower waste-disposal costs (or generate lower environmental costs) if production is sufficiently dispersed to allow natural absorption.

On the other hand, rising market share in developing countries for larger-scale producers of pork, eggs, and poultry meat suggests that other factors may be prevailing in determining who in developing countries is able to capture the fast-expanding market:

- Small farmers may not have the technical skills needed to produce these commodities. Even if they are familiar with producing chickens or milk for home consumption, the production techniques may be different for commercial production.
- Lack of credit or liquidity makes it difficult for small farmers to purchase specialized agricultural inputs or to make the investments needed to produce these commodities.
- Small farmers are less able than large ones to bear the risk associated with pro-

ducing highly perishable commodities. Not only do perishable commodities experience greater price fluctuations and risk of spoilage than do non-perishable ones, but once the commodity is ready for sale, perishability puts the farmer in a weak bargaining position relative to the buyer.

- Small farmers frequently do not have access to the information about market demand they need to make production decisions. The problem is not just lack of price information, but also lack of information about the relationship between price and product characteristics such as color, size, shape, texture, fat content, freshness, and so on.
- Buyers may not have access to information about the quality of output from specific smallholder farms, which makes them less willing to purchase from any smallholders at any given price level, compared to buying from a well-identified large-scale provider.
- Larger farms may be more able than smaller ones to secure policy subsidies such as subsidized credit or better infrastructure.
- Larger farms may be relatively more able than smaller farms to get away with creating negative externalities through pollution.

The presence of the latter factors suggests that rapidly rising demand for livestock products in developing countries may do more to stimulate the large-scale sectors than smallholder farmers. In fact, the rise of large-scale industrialized farms made possible by crossing some threshold of minimum assured demand could conceivably even lead to the reduction of smallholder sales below where they were before the demand surge.

Hypotheses to Be Tested
A literature review suggests that the main determinants of the hypothesized greater

competitiveness of large farms, and thus by assumption scaling-up, are

- increasing technical and allocative economies of scale in livestock, including economies of scale in input procurement;
- policy distortions such as scale-variant subsidies per farm;
- greater capture of negative environmental externalities per unit of output and land by larger farmers; and
- the role of posited higher transaction costs from asymmetries in access to information and assets facing small-holders in reducing their competitiveness relative to those of larger-scale farms.

Knowing which driving forces create the incentives for larger-scale livestock farms is critical to knowing what to do about it and whether to do so. The issues just cited can be clarified by the empirical testing of a number of specific hypotheses. Table 1.3 summarizes the links between testable hypotheses and inferences from test

results for what is driving incentives for the scaling-up of livestock farms.

If true economies of scale from technology, management, or transport (for example) are driving the incentives for larger scales of livestock farming, we would expect, ceteris paribus, larger farms both to be more profit-efficient and to have unit profits larger than or equal to those of small farms, where all variables are measured in financial rather than social prices. Such larger farms could eliminate competition from small family farms over time by cutting their profit margins. However, if small farms both are more profit-efficient and have unit profits higher than or equal to those of large farms, they are unlikely to be displaced by large farms, except possibly by events in supply chains occurring beyond the farm level, which is outside the scope of this study. On the other hand, if small farms are more profit-efficient than large ones but still have lower unit profits, either it is a temporary phenomenon or it reflects higher unit subsidies or other distortions bolstering the unit profits of large farms. Hence the rationale for two key hypotheses:

Table 1.3   Linking issues to expected empirical findings and tests

Issue investigated	Expected empirical findings	Test
Whether economies of scale (e.g., from technology) are contributing to scaling-up	Small farms have lower unit profits and lower relative profit efficiency	Hypotheses 1 and 2 are rejected
Whether policy distortions (subsidies to large-scale farms) are contributing to scaling-up	Small farms have lower unit profits *and* higher profit efficiency than large farms	Hypothesis 1 is rejected and there is a failure to reject Hypothesis 2
Whether environmental externalities are contributing to scaling-up	Small farms internalize more per unit of the cost of pollution from livestock production than do large farms	Failure to reject Hypothesis 3; failure to reject Hypothesis 4
Whether transaction costs are contributing to scaling-up	Transaction cost proxies have a greater negative effect on the relative profitability of small farms than do large ones	Failure to reject Hypothesis 5
Whether contract farming is more profit-efficient than independent farming	Contract farms have higher relative profit efficiency and unit profits than do independent ones at comparable levels of scale	Failure to reject Hypothesis 6
The relative importance of each driver in a simultaneous experiment	See Chapter 3	See Chapter 3

Hypothesis 1: Small-scale producers have higher or equal profits per unit of output than do large producers. This can be directly calculated from farm survey data, and it represents a benchmark. If supported, smallholders still have a chance to compete. If rejected, it is unlikely that smallholders will be able to compete in the same markets as larger-scale producers, because smallholders have, by definition, a small volume of output with which to make a living, and large farms will eventually drive small ones out of business. Even if supported, there is still the possibility that large-scale producers will drive smallholders out of the market by reducing prices if they are more efficient users of resources.

Hypothesis 2: Small-scale producers are more efficient users of farm resources to secure profits than are large-scale producers, other things equal. This issue requires an analytical methodology for empirical testing (see Chapter 3). By *efficiency* we mean the amalgam of technical efficiency in terms of producing the most valuable output with a given set of inputs and allocative efficiency, in terms of using the most efficient combination of inputs given prevailing prices.[1] If Hypothesis 1 and the present hypothesis are both supported, the outlook for smallholders is pretty good, because the combination of higher unit profits and greater efficiency would mean that the small farmers could either displace large farmers, or possibly they could eventually become large farmers!

If Hypothesis 1 holds and Hypothesis 2 does not, implicit subsidies to large-scale operations may be at work, and if so they may even overcome the efficiency advantage of small-scale production. If both hypotheses are rejected, there is little apparent scope for keeping smallholders involved without explicit subsidies for this purpose, and even then it is unlikely that their presence will be felt for long.

A spin-off subhypothesis here is that the relative profit efficiency of small farms compared to large ones depends greatly on how family labor input is valued in both cases. Because family labor is a much higher share of total labor on smaller farms, not valuing family labor favors the estimated relative efficiency of smallholders. In fact, consistent with the literature cited earlier, it seems likely that one way smallholders can avoid being displaced is by undervaluing their own labor relative to the market wage rates paid by commercial operators for similar work. If this turns out to be the basis of smallholders' competitiveness, the latter is likely to become more fragile as labor markets become more integrated over time.

There are other hypotheses to be tested as well.

Hypothesis 3: Small farmers expend a greater amount of effort/investment in abatement of negative environmental externalities per unit of output than do large farmers. This hypothesis requires an empirical approach for measuring externalities that can be attributed to specific farms (see Chapter 3). If such a methodology is feasible, it allows us to indirectly address a fascinating question: do large farmers reap more benefits per unit of output from environmental externalities than do small ones? If so, lack of enforcement of environmental laws is probably contributing to scaling-up. If not, smallholders may have been able to hang on in the livestock sector in part because they get away with pollution and larger-scale operators do not. However, increasing densities of smallholders producing livestock near human

---

[1]Kumbhakar (2001) shows that under neoclassical assumptions overall efficiency is the product of technical and allocative efficiency.

population centers under the Livestock Revolution will bring increasing problems.

Hypothesis 4: Uncompensated negative environmental externalities favor the nominal profit efficiency of large farms over small ones, ceteris paribus. If small farmers consistently internalize a larger share of negative externalities than do large farms, as in Hypothesis 3, it stands to reason that environmental externalities favor the nominal profit efficiency of large farms. However, a model is required to test Hypothesis 4, even if descriptive analysis is enough to support the assertion in Hypothesis 3. This is because the value of an uncompensated externality to the relative profit efficiency of a given farm is just one of many influences, and it may also be correlated with other relevant factors. Empirical methods used to assess the differential impacts of uncompensated externalities on relative profitability must also simultaneously account for all the other factors that explain differences in profitability across farms, especially differences in technical and allocative efficiency and differences in policy distortions and transaction costs as they affect different farms. If a satisfactory approach to these problems can be devised, the pay-off will be an unbiased view of the relative importance of uncompensated environmental externalities in promoting scaling-up.

Hypothesis 5: Profits of small-scale producers are more sensitive to "transaction costs" than are those of large-scale producers. The previous discussion suggested that transaction costs are likely to diminish the competitiveness of smaller farms more than that of larger ones. If that is true, it suggests that policies that reduce transaction costs proportionate

to output for all producers will be of particular benefit to smallholders. It also suggests that institutional solutions to high transaction costs, such as dairy cooperatives and contract farming, will be of the greatest benefit to smallholders. However, there are methodological problems in investigating this issue. First, transaction costs may be difficult to observe, and therefore must be inferred from farm characteristics that are likely to be collinear with differences in access to information and assets. Second, there is a problem similar to the one under the previous hypothesis involving sorting out different influences on relative profitability. As before, a satisfactory approach to these joint problems will yield an unbiased view of the relative importance of different kinds of information or asset asymmetry in promoting scaling-up.

Hypothesis 6: Contract farmers are more profit efficient than independent farmers at comparable scales of operation. If contract farming is a way around the high transaction costs plaguing smallholders, it should increase the efficiency with which they use their fixed farm resources to secure profits. This can be expected to also show up in higher average unit profits for contract farmers, although the latter will vary from year to year depending on market price fluctuations. If profit efficiency and unit profits are higher for contract farmers, ceteris paribus, contract farming, where it is appropriate, could keep small-scale farms participating in high-value markets for some time to come. From a policy perspective, it will also be important to know if the higher profitability of contract farming is due in part to implicit or hidden subsidies received by the integrating institution.

CHAPTER 2

Growth, Concentration, and Integration of the Livestock Sector in the Study Countries

T he four country cases chosen—the Philippines, Thailand, India, and Brazil, going from east to west—have diverse histories, economies, cultures, and political structures. However, they are all fast-changing developing countries where cities, population, urban incomes, and consumption of livestock products have been growing rapidly since the early 1980s. They have also all seen, at least in the initial stage of livestock industrialization, the rise of large periurban livestock operations of one form or another, typically not far from major cities. Finally, they also typically have vibrant smallholder livestock sectors producing similar products nearby.

Growth and Concentration in the Philippines

The livestock sector provided the strongest source of agricultural growth as the Philippine economy emerged from recession in 1998.[1] In particular, pork and broiler meat have been at the forefront of the livestock production growth, growing at 3.2 and 7.2 percent per year in 1990 and 2005, respectively. This growth is driven mostly by a surge in demand, an increase in imports of breeding stock, the use of vaccines and drugs to control for diseases, and vertical integration in the case of poultry. Expansion of these two industries has been concentrated in the Central Luzon and Southern Tagalog regions, located north and south, respectively, of the capital city of Manila, the largest demand center (see http://www.fao.org/WAIRDOCS/LEAD/X6170E/x6170e0c.htm for the geographic concentration of swine and poultry in the Philippines). Both regions have relatively increasing incomes and expanding populations and thus are also centers for growth in consumption, in addition to the nearby Manila market.

Chicken farms in the Philippines were initially characterized by the use of native breeds (Figure 2.1).[2] Native chickens are mostly raised on backyard farms by smallholders,[3] and broilers (which are imported hybrids raised using free-range practices) were referred to as "commercial" chickens by the Bureau of Agricultural Statistics until 1998. Feed consists mainly of crop residues and grain spillage along with broken rice and corn. Native breeds con-

---

[1]This section is drawn from Costales et al. (2003), which gives appropriate citations to original sources.

[2]Native chickens are those that do not include the recently imported hybrid chickens with foreign strains and include improved breeds (those that are crosses of local chickens with foreign strains).

[3]A poultry farm in the Philippines is classified as "backyard" if it has fewer than 100 birds (Philippines, BAS 1987).

Figure 2.1   Native chicken and broiler chicken inventories, Philippines, 1990–2006

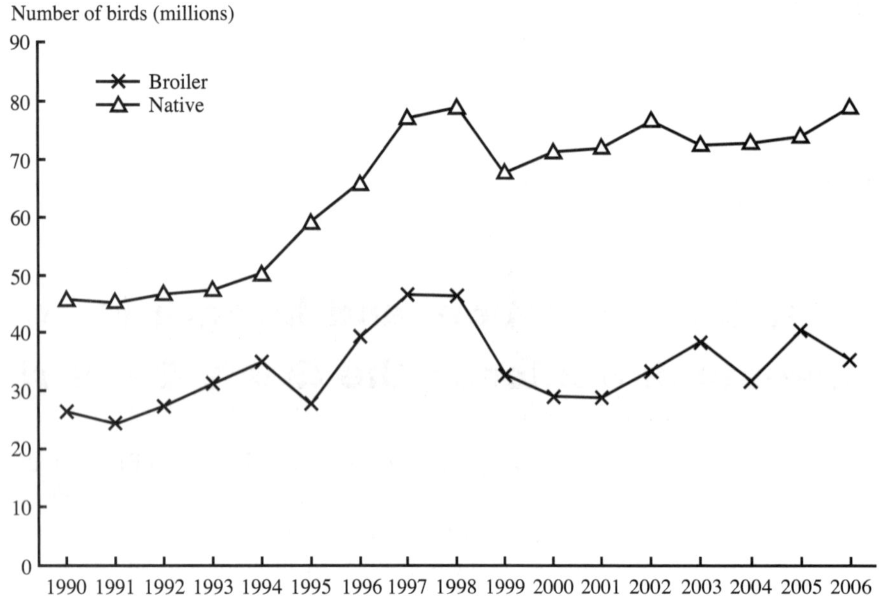

Number of birds (millions)

Sources: Costales et al. (2003); Philippines, Bureau of Agricultural Statistics (BAS) (2007a).

tinue to be important in the Philippine poultry market, and the number of native chickens relative to commercial chickens seems to have increased, while commercial growth has remained pretty constant despite the short peak between 1995 and 1998. However, Babcock-based commercial breeds are rapidly displacing them in the growing Metro Manila market and nearby major market areas in Central and Southern Luzon (Figure 2.2).

Although smallholder hog and poultry production at all scales has grown nationwide, large-scale industrialized livestock production has grown even faster, particularly in Central and Southern Luzon. The broiler sector has been at the forefront of livestock industrialization, from importing grandparent stock and medicines and using advanced technologies to promote growth to developing vertically integrated production systems through contract farming. The latter span the range from feed ingredient importing to retailing processes meat.

Large integrators control some 80 percent of the broiler market in the Philippines.

The remainder is in the hands of independent commercial broiler producers, who typically maintain inventories of 20,000–100,000 birds, and a few smallholders, who typically raise 1,000–2,000 birds at a time. The large integrators are banded together into an organization, the Philippine Association of Broiler Integrators, consisting of six large firms. The integrators engage in breeding and contract growing, processing, and distribution of branded output. They are also major players in the live broiler market, in which the daily reference price is set by the group in the form of a "gentlemen's agreement" (Costales et al. 2003). The integrators are also the main source of day-old chicks for independent commercial producers and smallholders.

With regard to feed supply, while small-sized feed mills have their own brands of mixed feeds, the integrators not only supply their own feeds for internal use, but also compete with small-scale mills in the commercial market as suppliers of mixed feeds. Anecdotal evidence suggests that the integrators have relatively easier access to mini-

Figure 2.2   Broilers (imported hybrids) dominate in the Metro Manila area (Regions III and IV-A), 2006

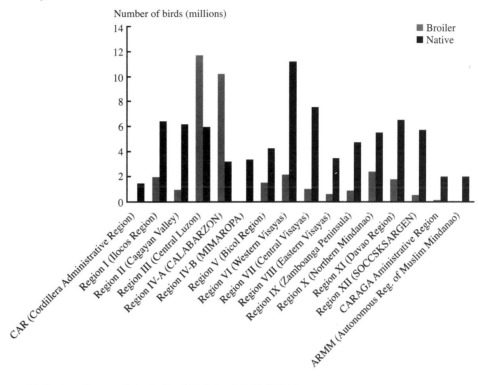

Source: Philippines, Bureau of Agricultural Statistics (BAS) (2007a).

mum access volume guarantee corn imports negotiated at 35 percent under the General Agreement on Tariffs and Trade (Costales et al. 2003). They are thus better able to compete against other popular feed brands based on domestic corn prices that reflect the prevailing corn tariff of 60 percent. It is also possible that being horizontally integrated with grain mills for human food use conveys economies of scope in feed grain importation and milling.

The live broiler market is still the main output market for smallholders and large independent commercial broiler producers because of the continued preference of consumers for fresh meat. The large integrators control the market for dressed broilers, which they sell through vertically coordinated retail meat shops and an emerging urban supermarket sector. In an era of market uncertainties created by international

trade in meat, independent producers have less capacity to cushion their incomes from relatively large fluctuations in live broiler prices as compared to dressed meat prices.

In the hog sector, which is less concentrated than the broiler industry, commercial farm operations have also concentrated in the two provinces surrounding Metro Manila. Small, medium-sized, and large-scale producers have relatively equal access to the major market, the live market for slaughter hogs served by small-scale collectors who truck small butchers of live animals to slaughter points in Metro Manila.

The truly larger-scale commercial firms in the hog sector are vertically integrated, from breeding and contract production to slaughter and processing of branded meat products. These large firms increasingly access a higher-value retail market through the growing institutional sector in large cities, which includes supermarkets and up-scale restaurants and hotels, than do the medium-sized commercial and smallholder farms. Through vertical coordination, these large firms can meet the food safety certification needs and consistent quality standards of the institutional food sector. However, vertical integration and contract growing have not yet become the norms throughout the hog industry. Currently, no one integrator has a dominant market share, unlike in the chicken industry.

In sum, livestock production in the Philippines has grown on both small (backyard) and large-scale (commercial) farms and has become highly concentrated around major demand centers. In fact, large-scale farms have overtaken backyard farms in proximity to Manila (Region IV-A),[4] as in the case of hog production (see Figure 2.3). Large-scale layer farms have already completely displaced smallholders, and large-scale broiler operations are starting to do the same.

Growth and Concentration in Thailand

The beginnings of structural change in livestock production in Thailand can be traced to the introduction of modern production methods in 1946 and the importation of high-yielding stock in 1956 (Poapongsakorn 1985).[5] Yet major changes did not occur until local and regional markets provided an incentive for major production increases with the start of the Livestock Revolution in the 1970s. At that time, the Charoen Pokphand Company (CP) established its chicken-breeding business through a joint venture with the U.S.-based Arbor Acres Company, bringing improved grandparent stock into Thailand. Subsequently CP went on to become a diversified and vertically and horizontally integrated multi-national corporation rooted in the feed, meat, and shrimp trade, with $4 billion U.S. dollars in sales revenues and 100,000 employees by the mid-1990s.

With extension of the CP model to poultry producers, Thai broiler chicken production went from 474,000 tons in 1983 to over 1.2 million tons in 2001 (Poapongsakorn et al. 2003). Such rapid growth of broiler production was possible because of the export markets made accessible through unit cost-cutting technology and organization put in place by the private sector and through supportive regulation and trade liberalization by the state. Sharply lower chicken prices and positive income growth were also associated with a doubling of domestic chicken consumption, from 6.8 kg per person in the early 1980s to over 12 kg in 2001. Rapid expansion of Asian regional demand and European demand in the 1990s provided a vent for expanding production. The Thai broiler market is now dominated by a dozen integrators who control the complete supply chain, ranging from grandparent stock farms to the food retail business and export trade.

Egg consumption and the layer industry have also expanded significantly since the early 1980s, although not as impressively as the growth in broilers; roughly 90 percent of eggs produced are for domestic consumption. The annual domestic consumption per capita jumped from 66 eggs in 1983 to 145 eggs in 1995, then dropped slightly, to 139 eggs in 2001. The supply-side growth of the layer industry can also be explained by the similar factors affecting the broiler industry: the adoption of modern technology and rapid economic growth in the late 1980s. New technology has enabled the industry to accelerate its egg production, from nearly 3.3 billion eggs in 1983 to almost triple this figure (9 billion) in 2001. Although poultry production has grown rapidly in Thailand in the last

---

[4]Region IV-A consists of the provinces of Cavite, Laguna, Batangas, Rizal, and Quezon.

[5]This section draws on Poapongsakorn et al. (2003), which gives appropriate citations for original sources.

Figure 2.3   Growth of commercial and backyard inventories on hog farms in Southern Luzon (Region IV-A), 1990–2006

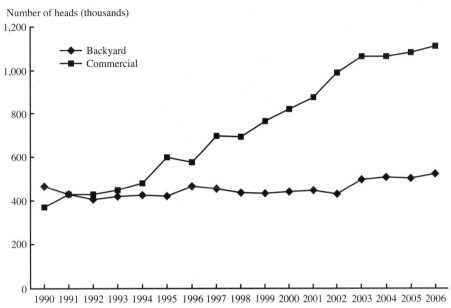

Source: Philippines, Bureau of Agricultural Statistics (BAS) (2007b).

several decades, the recent avian influenza has brought devastation to the industry.

Although modern swine breeds were first introduced in the 1960s and programs were put in place in the mid-1970s to train farmers in modern production and farm management, major changes in the Thai swine industry began to take place only in the mid-1980s. The introduction of exotic pig breeds and evaporation shed cooling has begun to move Thailand into industrial swine production. The production of hogs increased from 11 million head in 1983 to more than 16 million in 2001. Between 1988 and 2003, the share of Thai swine farms with holdings of more than 100 sows went from 0.3 percent to 2.0 percent. Between 1993 and 2003 alone, the share of all pigs on farms with holdings of fewer than 20 sows went from 55 percent to 21 percent (Table 2.1). The pace of expansion has been slowed by disease problems that limit exports to high-value markets, by the restrictive regulation of slaughterhouses, feed costs, and by burgeoning environmental concerns.

Since the mid-1990s, the Thai government has discouraged the creation of new large-scale farms in the three largest pig-producing provinces, Nakorn Pathom in the west and Chachoengsao and Cholburi in the east, because of land constraints and the high risk of spreading disease in areas with an already high animal density (see http://www.fao.org/WAIRDOCS/LEAD/X6170E/x6170e0c.htm for the geographic concentration of dairy, poultry, and swine in Thailand). In effect, new large-scale pig farms expanded into Ratchaburi (in the west) and Saraburi and Lopburi (in the east). These provinces, particularly the latter two, have abundant water resources. Increasing water pollution caused by the farms situated near the extensive river system in Nakorn Pathom and Chachoengsao is becoming a growing concern.

Unlike the case of poultry and swine development, which has been driven by the private sector, the government has driven dairy sector development in Thailand, with the goal of developing small-scale

Table 2.1 Farm size in Thailand by type of animal, selected years 1978–2003

Number of swine in holding	Percentage share in number of swine					Percentage share in number of swine in holding				
	1978	1988	1993	1998	2003	1978	1988	1993	1998	2003
1–4	n.a.	n.a.	25.26	10.07	5.58	79.35	80.15	21.16	60.66	53.27
5–19	n.a.	n.a.	29.89	22.34	15.4	17.83	16.46	39.52	30.1	34.36
20–49	n.a.	n.a.	20.8	16.72	11.36	2.29	2.51	25.5	7.29	8.62
50–99	n.a.	n.a.	8.28	4.24	5.05	0.35	0.54	8.6	0.82	1.69
100 or more	n.a.	n.a.	15.82	46.63	62.61	0.18	0.34	5.22	1.13	2.06
Total (percent)	n.a.	n.a.	100.00	100.00	100.00	100.00	100.00	100.00	100.00	100.00
Total (number of swine and farmers)	5,314,000	4,684,926	6,182,953	5,731,360	7,182,512	1,263,000	778,113	590,616	465,509	344,322

Number of dairy cows in holding	Percentage share in number of dairy cows					Percentage share in number of dairy cows in holding				
	1978	1988	1993	1998	2003	1978	1988	1993	1998	2003
1–4	n.a.	n.a.	9.21	5.76	4.50	n.a.	54.73	11.88	34.67	28.59
5–19	n.a.	n.a.	44.04	48.90	35.62	n.a.	39.85	39.15	49.56	46.70
20–49	n.a.	n.a.	26.66	34.90	44.08	n.a.	4.53	31.42	15.07	22.14
50 or more	n.a.	n.a.	20.09	10.44	15.82	n.a.	0.89	17.54	0.69	2.56
Total (percent)	n.a.	n.a.	100.00	100.00	100.00	n.a.	100.00	100.00	100.00	100.00
Total (number of dairy cows and farmers)	n.a.	93,654	257,895	240,206	446,455	n.a.	15,966	11,616	21,540	31,404

Number of chickens in holding	Percentage share in number of chickens					Percentage share in number of chickens in holding				
	1978	1988	1993	1998	2003	1978	1988	1993	1998	2003
1–19	n.a.	n.a.	9.15	10.15	1.29	69.30	69.77	64.24	61.35	35.17
20–99	n.a.	n.a.	16.91	20.15	7.11	30.00	29.11	33.00	36.09	56.46
100–499	n.a.	n.a.	4.80	4.96	3.12	0.53	0.84	2.03	1.99	6.41
500–999	n.a.	n.a.	1.54	1.00	0.40	0.05	0.00	0.15	0.08	0.18
1,000–9,999	n.a.	n.a.	29.11	31.16	30.80	0.09	0.28	0.50	0.42	1.38
10,000 or more	n.a.	n.a.	38.49	32.58	57.28	0.03	0.00	0.09	0.06	0.39
Total (percent)	n.a.	n.a.	100.00	100.00	100.00	100.00	100.00	100.00	100.00	100.00
Total (number of chickens and farmers)	54,157,000	86,679,292	154,921,930	169,102,499	248,982,983	2,638,000	3,249,177	2,617,412	3,174,410	1,028,189

Sources: Poapongsakorn et al. (2003), citing Thailand, National Statistical Office (1988, 1993, 1998); Thailand, National Statistical Office (2003).
Note: n.a., not available.

production. The government has directly imported dairy breeding stock, provided production subsidies and tariff and quota protection, and played a coordinating role between dairy producer cooperatives and dairy processing companies. Through the aid of coordination and import restrictions, farmer cooperatives have been able to sell all their raw milk at much higher than world prices.

During 1993–95, the growth of the dairy industry was stimulated by the government's project to provide free milk to children in elementary schools for 260 days a year. Subsidized milk for students must come only from domestic milk production; it currently accounts for half of domestic milk consumption. Domestic milk consumption in 2001 was 9 kg of milk per person, compared to 1 kg in the early 1980s.

In sum, the Livestock Revolution in Thailand has occurred only in the last 20 years, beginning with the export-led broiler sector. Although private sector innovations such as improved breeds, feed technology, housing, farm management, and contractual arrangements between integrators and farmers have been the prime sustainers of growth, export opportunities and rapid domestic and regional economic growth during 1985–95 were the essential catalysts. Swine and dairy development have been driven by domestic market demand and have been significantly affected by policy factors—regulation of slaughterhouses for swine, subsidies and protection for dairy. The livestock industry has grown in close proximity to Bangkok, and heavy concentrations of animals are causing environmental stress. Farm sizes have become significantly larger, even in the case of dairy farms, as shown in Table 2.1. However, this does not necessarily mean that smallholders have experienced increased poverty. Today's large farmer in Thailand may have

been a smallholder 20 years ago, and many in the dynamic central region (where livestock production is concentrated) were. Besides, the fact that small farms have exited from the sector may mean only that they found something more profitable to do in the area of either high-value agriculture or nonfarm work.

Available information from secondary sources in Thailand indicates that the percentage of poor livestock households has declined proportionately more than the percentage of poor agricultural households overall (Table 2.2 and Poapongsakorn et al. 2003). Moreover, the Gini coefficient for livestock farmers is also much lower than that for agriculture as a whole (Table 2.2 and Poapongsakorn et al. 2003). This suggests that the present income distribution among the livestock-keeping households is relatively more equitable.[6]

Comparison of data from the 1993 Agricultural Census and the Inter-censal Survey of Agriculture in 1998, reported in Table 2.3, shows that the average amount of land used for growing crops and rearing livestock increased from 1993 to 1998, and it is likely that the livestock holdings per farm also increased. However, this does not suggest a clear relationship between the amount of land used for growing crops and rearing livestock and income growth over the period. Furthermore, the real income of livestock-producing households increased between 1993 and 1998, despite the fact that 1998 was the worst year of the economic depression that resulted from the Asian economic crisis, when the GDP growth rate was a negative 8 percent. However, for livestock producers, the farms with the greatest growth of income in this period were the smallest ones in terms of land size; farms on less than 1 hectare of land saw their real livestock incomes grow at 6 percent per annum in 1994 domestic prices.

---

[6] A Gini coefficient is tantamount to a cumulative average of income inequality by class, where a perfectly equal distribution across classes would yield a Gini of 0 and a perfectly unequal distribution would score 1.

Table 2.2 Percentage of livestock households and agricultural households and their farm income, 1986–2000

Year	Percentage of households in livestock or agriculture			Percentage of household income from farming[a]				Gini coefficient ratio		
	Livestock versus agricultural[b]	Poor livestock versus all livestock[c]	Poor agricultural versus all agricultural[d]	Agricultural households	Livestock households	Poor agricultural households	Poor livestock households	All households	Agricultural households	Livestock households
1986	0.4	30.0	46.3	21.8	9.4	19.6	12.3	0.50	0.32	0.19
1988	0.6	33.8	41.2	31.6	26.0	23.6	23.7	0.90	0.34	0.57
1990	0.6	19.8	34.1	36.2	29.7	22.6	19.9	0.51	0.34	0.31
1992	1.0	18.4	31.6	26.4	12.4	17.8	11.1	0.54	0.33	0.21
1994	0.4	2.0	25.1	20.9	5.6	12.1	0.0	0.52	0.34	0.13
1996	0.5	13.9	17.4	23.2	8.5	11.4	0.4	0.52	0.35	0.25
1998	0.5	5.7	17.3	26.1	16.2	9.2	0.0	0.51	0.36	0.22
1999	0.6	n.a.	23.1	26.9	19.6	17.7	n.a.	0.53	0.36	0.18
2000	1.4	1.4	22.7	50.2	58.6	12.5	8.3	0.53	0.37	0.52

Source: Poapongsakorn et al. (2003), calculated from National Statistical Office, socioeconomic survey data tape.
Notes: The 1986–99 definition of livestock occupation is livestock workers, dairy farm workers, and poultry farm workers. The 2000 definition of livestock occupation is market-oriented animal producers and related workers. n.a., not available.
[a]Values calculated as household farm income divided by total household income.
[b]Values calculated as the number of households engaging in livestock raising divided by the number of households engaging in agriculture.
[c]Values calculated as the number of poor households engaging in livestock raising divided by the total number of households engaging in livestock.
[d]Values calculated as the number of poor households engaging in agriculture divided by the total number of households engaging in agriculture.

Table 2.3 Household income from agricultural produce and number of farm holdings, 1993 and 1998 (current and 1994 baht)

Type of household	Average income at current prices		Average income at 1994 prices		Average growth rate (percent)
	1993	1998	1993	1998	
	Household income by farm area (rais[a])				
Holders rearing livestock only	204,747	289,345	215,524	227,831	1.1
Fewer than 6	21,545	39,552	22,679	31,143	6.3
6–9	31,149	39,601	32,788	31,182	–1.0
10–39	33,265	47,529	35,016	37,424	1.3
40–139	51,828	65,329	54,555	51,440	–1.2
140 or more	66,961	97,334	70,485	76,641	1.7
Holders growing crops and rearing livestock	162,193	225,512	170,729	177,569	0.8
Fewer than 6	13,542	18,675	14,254	14,704	0.6
6–9	14,332	20,877	15,086	16,439	1.7
10–39	20,059	35,634	21,115	28,058	5.7
40–139	39,035	61,442	41,089	48,380	3.3
140 or more	75,225	88,884	79,184	69,987	–2.5
Holders rearing livestock only	Number of holdings by annual income (baht[b])				
Fewer than 5,001	4.7	1.1			
5,001–10,000	9.2	2.8			
10,001–20,000	13.7	8.3			
20,001–50,000	27.1	15.6			
50,001–100,000	24.4	33.0			
100,001 or more	21.0	39.2			
Total	100	100			
	Holders growing crops and rearing livestock				
Fewer than 5,001	3.6	0.8			
5,001–10,000	9.9	3.7			
10,001–20,000	18.0	10.7			
20,001–50,000	36.5	35.9			
50,001–100,000	21.0	31.3			
100,001 or more	11.0	17.6			
Total	100	100			

Source: Poapongsakorn et al. (2003), citing Thailand, National Statistical Office (1993, 1998).
[a]One hectare is equal to 6.25 rais.
[b]In 1998, US$1 was approximately equal to 36.6 baht; in 1993 it was approximately equal to 25.6 baht.

Growth and Concentration in India

Red meat consumption in India is very low, primarily for cultural and religious reasons.[7] But India has participated in the global Livestock Revolution through extraordinary growth in the consumption of milk, eggs, and poultry meat. In fact, poultry is one of the fastest-growing segments of the agricultural sector in India today. While the production of crops has been rising at a rate of 1.5–2 percent per annum over the past 20 years, production of eggs and broilers has been rising at a rate of 8–10 percent per annum. As a result, India is now the world's 5th-largest egg producer and the 18th-largest producer of broilers (Mehta et al. 2003).

A significant feature of India's poultry industry has been its transformation from a backyard activity into a major commercial activity in four decades. This transformation has involved sizeable private sector investments in breeding, hatching, rearing, and processing. Farmers in India have moved from rearing indigenous breeds to producing internationally recognized hybrids. With the exception of sterile egg powders exported to Europe, virtually all Indian poultry and egg production is consumed in South Asia, most of it in India.

As in Thailand, India's growth in the poultry industry began with the importation in 1974 of grandparent stock from Cobb, a dominant multinational firm at the time. Cobb and Venkateshwara Hatcheries set up a joint venture in the 1980s to produce pure-line parent stock. Venkateshwara now exports breeding stock to Bangladesh, Bhutan, and Nepal.

Four southern states—Andhra Pradesh, Karnataka, Kerala, and Tamil Nadu—accounted for about 45 percent of the country's egg production, with per capita annual consumption of 57 eggs and 0.5 kg of broiler meat in 2001. Production is grouped around major urban centers such as Hyderabad, Mumbai, and Delhi (see http://www.fao.org/WAIRDOCS/LEAD/X6170E/x6170e0c.htm for the geographic concentration of poultry in India). The eastern and central regions of India accounted for roughly 20 percent of egg production; per capita consumption was 18 eggs and 0.13 kg of broiler meat nationally in 2001. The northern and western regions of the country record much higher per capita production of eggs and broiler meat.

The growth of the poultry sector in India is marked by an increase in the size of poultry farms. In earlier years, broiler farms produced on average a few hundred birds (200–500 chicks) per cycle. Today, units with fewer than 5,000 birds are rare, and units with 5,000–50,000 birds per cycle are common. Similarly, layer farms with a flock size of 10,000–50,000 birds have become common. Small units are probably finding themselves at a disadvantage because of higher feed and transport costs, more expensive vaccines and veterinary care services, and the nonavailability of credit.

Despite India's growth in consumption, its per capita consumption of these products is poor: 37 eggs and 1 kg of poultry meat per capita per annum. There is considerable variation in per capita consumption of poultry products between rural and urban areas and across the region. For instance, per capita consumption of eggs is only 7.7 per annum in rural areas, compared with 17.8 per annum in urban areas, and in seven states, per capita consumption is less than 3.5 per annum. Similarly, per capita consumption of poultry meat is 0.24 kg in rural areas and 1.08 kg in urban areas (Mehta et al. 2003).

The performance of the Indian dairy sector during the past three decades has also been impressive. Milk production grew at an average annual rate of 4.6 percent during

---

[7]This section draws on Mehta et al. (2003) and Sharma et al. (2003), which give appropriate citations for original sources.

the 1970s, 5.7 percent during the 1980s, and 4.2 percent during the 1990s. Until 1991, the Indian dairy industry was highly regulated and highly protected. Milk processing and dairy product manufacture were restricted mainly to small firms and cooperatives. High import duties, nontariff barriers, restrictions on imports and exports, and stringent licensing provisions provided incentives to Indian-owned small enterprises and cooperatives to expand production in a protected market. In March 2002, the delineation of milk-shed supply areas, which gave preferential access to milk supplies to cooperatives, was abolished. This resulted in an increase in the private dairy-processing plants, from 248 in 1991 to 516 in 2003/4, but the informal sector remains dominant in the milk markets (Birthal and Taneja 2006).

Several factors have contributed to increased milk production. First, milk and dairy products have cultural significance in the Indian diet. A large portion of the population is lacto-vegetarian, so milk and dairy products are an important source of protein in the diet. The demand for milk and dairy products is income-responsive, and growth in per capita income is expected to increase the demand for milk and milk products. The growth was also achieved through the intervention of the Indian government. Producing milk in rural areas through smallholder producer cooperatives and moving industrially processed milk from these smallholder sources to urban demand centers became the cornerstone of government dairy development policy. This policy initiative gave a boost to dairy development and initiated the process of establishing the much-needed linkages between rural producers and urban consumers.

Despite being the largest milk producer in the world (producing 84.6 million tons of milk in 2001–2), India's per capita availability of milk is low by global standards,

although it is high by developing country standards. The per capita availability of milk, which declined during the 1950s and 1960s (from 124 grams per day in 1950–51 to 121 grams in 1973–74) expanded substantially during the 1980s and 1990s, reaching 226 grams per day in 2001–2. Per capita consumption of milk and milk products in India is among the highest in Asia and is still growing. It is still below the world average of 285 grams per day, and also the minimum nutritional requirement of 280 grams per day as recommended by the Indian Council of Medical Research (Sharma et al. 2003).

Even though milk production in India is widespread throughout the country and overwhelmingly carried out by small-scale producers, there are still large interregional and interstate variations in milk production (see http://www.fao.org/WAIRDOCS/ LEAD/X6170E/x6170e0c.htm for the geographic concentration of milk production in India). Roughly two-thirds of national milk production comes from the states of Uttar Pradesh, Punjab, Rajasthan, Madhya Pradesh, Maharashtra, Gujarat, Andhra Pradesh, and Haryana. The eastern region is lagging behind in terms of dairy development and imports milk from surplus-milk-producing areas in the west and north.

Growth and Concentration in Brazil

Poultry and swine production have become major commercial activities over the last two decades in Brazil, propelled by technology and rapid development of the feed sector.[8] Poultry production used to be primarily located on smaller farms in the south; in the mid-1990s, a typical farmer in Santa Catarina province would have 6,000–15,000 broilers. Growth has expanded since then, with the new investments in the center-west on a much larger scale (more than 10,000

---

[8]This section draws on Camargo Barros et al. (2003), which gives appropriate citations for original sources.

birds per cycle) and more specialized farms (see http://www.fao.org/WAIRDOCS/LEAD/X6170E/x6170e0c.htm for the geographic concentration of poultry in Brazil). Poultry production continued to grow due to gains in productivity from improved breeding stock and firm demand from the export market as well as the domestic market (Camargo Barros et al. 2003).

However, in the south the average poultry farm may in fact have declined further in size over the past 15 years as land and environmental pressures have come to bear and as integrators have put pressure on small-scale producers to increase the size of their operations to lower their integrator costs in supplying inputs, technical assistance, and picking up animals for slaughter. Farmers who took out loans to expand their operations in the south found it difficult to withstand the devaluation of the real in the 1990s, and many of those interviewed who did scale up by taking out loans felt others (who were not interviewed) went bankrupt due to the escalating prices of inputs necessary to maintain the increased number of animals they were now raising. Although this may have been the rationale in reality, one would have thought that this would have been profit-neutral. Future analysis should look at the reason for this perception. Rapid rises in feed costs since 2001 are also contributing to the rapid exit of small- and medium-scale producers in the south who cannot make the move to the center-west.

Similar trends have been observed in the swine industry. The center-west of Brazil is rapidly evolving into one of the lowest-cost swine production zones in the world. Production of swine, as in the cases of poultry and milk, is following feed sources to the extensive farming belt of the center-west (see http://www.fao.org/WAIRDOCS/LEAD/X6170E/x6170e0c.htm for the geographic concentration of swine in Brazil). Casual observation also suggests that many small-scale swine farms exited in the south after the establishment of increased sanitary controls in the mid-1990s in connection with the control of foot-and-mouth (FMD) disease.

The development of milk production has been marked by a shift from nonspecialized farms where milk was almost a byproduct to specialized and vertically integrated milk production. The transition was spurred by liberalization of pricing in 1991 after decades of cooperative dairy marketing under tight government regulation of the sector. Since the mid-1990s, private processors have enforced the farmers's adoption of on-farm chilling and other technologies that have made continued participation by producers of less than 100 liters of milk per day infeasible. Small dairy farming has traditionally been located in the temperate south, in states such as Santa Catarina; in 1996, 28 percent of farms in the south had more than 70 cows, up from 18 percent in 1985. Yet the main growth area has been in center-west, in settlement areas of states such as Goias, home to new and large farms (see http://www.fao.org/WAIRDOCS/LEAD/X6170E/x6170e0c.htm for the geographic concentration of milk production in Brazil). In 1996, 81 percent of farms in the center-west had more than 70 cows, up from 69 percent in 1985 (Camargo Barros et al. 2003).

There has been shrinkage in the total number of Brazilian milk producers due to greater competition with imported milk and new chilling requirements imposed by processors following liberalization of cooperative regulation and pricing in the 1990s. In 2000, there were 123,000 milk producers in Brazil that were producing 6 million liters delivered to the 15 largest processors. In 2002, the number of producers had shrunk to 95,000 that were producing the same total volume of milk (Camargo Barros et al. 2003).

It is difficult to separate livestock development in Brazil from three major structural changes that occurred in the early 1990s: liberalized internal markets in combination with a move toward an outward-looking orientation in agriculture; achievement of animal disease control objectives in part of Brazil for export without vaccination to the countries of the Organisation for Economic Co-operation and Development; and the phenomenal growth of the feed-grain industry. Of the four countries studied, Brazil is the country in which the smallholder is probably least likely to be able to survive as an independent producer, because Brazil has already moved in most cases, with the exception of dairy, into an industrialized production process reaping the benefits of scale, sourcing inputs largely from medium to large producers to reduce the transaction costs associated with the purchasing of inputs and marketing of outputs.

Given the ubiquitous trend of growth of production combined with increased geographic concentration and larger-scales of operation observed in all the study countries, it is interesting to consider an institutional change that has occurred in all four countries that has permitted some smaller-scale farms to make a smoother transition. This is contract farming. It explicitly reduces the transaction cost of selling or procuring a specific volume of output each day due to the improved communication on quality and price information implicit in such an arrangement.

Vertical Integration of Smallholders through Contract Farming

Modern forms of contract farming in milk production have existed in Europe since at least the mid-19th century; the traditional Danish dairy co-op was in fact a form of contract farming. The contract is an institution that reduces the pernicious effects of information and asset asymmetries across market actors, especially for smallholders and those who deal with them. It allows all parties to reduce the resources they need to search, monitor, and control quality when selling or buying a perishable product in uncertain or thin markets, and in this case al-

lows producers to obtain higher prices from a buyer who is fairly certain that the farmer will deliver clean, fresh milk on time. The institution also shares risks and captures economies of scale through the bulk purchasing of inputs. Dairy cooperatives in Brazil, India, and Thailand have all taken slightly different forms, but they share the common advantage of leaving more wealth to share between producers and processors through the reduction of transaction costs that are a net loss to producers, processors, and consumers combined. While it is clear that the cooperative mode will reduce transaction costs and that transaction costs are especially high for smallholders, it is unclear if the latter would remain involved without a government subsidy of some form.

More recently, contract farming has appeared in all the study countries in the swine and poultry sectors. Within the latter, contracts are more common in the production of broilers than of eggs, but they are observed in egg production in India in a few cases where industrial processors require quality control of inputs for specialized industrial outputs (Mehta et al. 2003).

Contracts observed in the country studies differed somewhat across countries and commodities. Forward price contracts for Indian broilers are more informal than those in the Philippines, where in turn they are more informal than in Brazil, for example. A contract growing arrangement in broiler and hog production is generally a contract between an "integrator," who supplies the intermediate inputs and procures the output, and a grower, who provides the primary inputs in the production process. The integrator provides the growing stock (day-old chicks, fatteners), feeds, and veterinary supplies and services and implements the final marketing of the output. The contract grower typically provides the space and facilities (land and housing), manure and dead animal disposal, equipment, utilities, labor (family and/or hired), and day-to-day farm management and deals with the neighbors and local authorities, including on environmental issues.

The form of contract across countries, regions, and commodities tends to be driven by four things. First, the changing needs of markets require changing product attributes, and these changing attributes (such as food safety) may not be easily observed at the time of sale. Contracting may permit processors a higher degree of quality control under these circumstances than employer-employee relationships would do. Second, different commodities embody different types of transaction cost and thus require different forms of institutional solutions. The information asymmetries between market participants in milk sales are fundamentally different than those in swine sales, for example. Third, contract farming is a sharing of risks and benefits between seller and buyer. As such, the precise form it takes depends greatly on the distribution of power (market and political) between buyers and sellers, as does enforcement of contracts. Fourth, integrators who contract production to a large number of smallholders are more likely to decrease their risk of being held accountable for environmental pollution than is a single large-scale production operation. The latter issue has probably had more of an effect on contracting in the developed countries than in the developing ones, but is already becoming an issue in Southeast Asia.

CHAPTER 3

Approach

The empirical approach to analysis of the scaling-up of livestock production is to define a quantitative measure of relative farm competitiveness in production, then look in a structured way at all the factors that differ across farms that might explain higher relative competitiveness. These factors include technical and allocative efficiency, scale economies positively associated with productivity (such as cutting overhead costs), asymmetries in access to assets (credit, liquidity, fixed capital, and so on) and information (education, experience, and such), externalities (some farmers get away with uncompensated pollution while others do not), and policies (some get a better deal from the government than others do).

The omission of relevant factors leads to biased estimates. Furthermore, inclusion of explanatory factors of relative competitiveness that are themselves functions of relative competitiveness leads to simultaneity bias. For example, relative competitiveness might be enhanced by being recognized as a sales leader, but being recognized as a sales leader may depend on being more competitive than others. The two-way causality among the variables leads to bias in the empirical estimation of the effect of all variables unless appropriate procedures are used.

Defining a Farm-Specific Measure of Relative Competitiveness

Relative competitiveness might be thought of as having the ability to produce at a lower unit cost of production than one's competitors. In fact, if large farms can produce livestock at a lower unit cost than small ones, they will clearly drive small farms out of the market over time. The market price that applies to both large and small farms, by this reasoning, will fall as large-scale producers expand production, and the smallholders will be squeezed out. The only future for smallholders, then, will be to stay in a few higher-priced niche markets not served by larger farms, if these markets exist, and to cut costs by paying less for their own (family) labor than what a large farmer pays for hired labor. Even so, it is unlikely that smaller producers will be able to stay in business long in this situation.

However, the reverse is not necessarily true. If small farms can produce at lower unit costs than large farms, they may still be squeezed out. This is because large-scale farms can remain profitable with very thin profit margins; they make up in volume what they lose in per-unit profit. Very low per-unit profits coupled with a small sales volume may not provide enough income for a smallholder to stay in business, at least not by specializing in livestock. Thus, if large farms have lower per-unit costs of production when all labor is costed at market wages, the next question is whether this finding still holds if smallholders do not cost their family labor. If it does, it is not necessary to proceed further; there is little hope for smallholders who wish to undertake this activity, except perhaps as a hobby or for minor income supplementation.

If, on the other hand, small farms can produce at lower per-unit cost in the same markets as large farms, perhaps by not costing their own labor at full market wage rates or for some other reason, there is at least some hope for them. Thus, having a higher per-unit profit, with or without the cost of family labor, is a necessary condition for the competitiveness of smallholders, but is not a sufficient one.

To get a more satisfactory measure of relative competitiveness that gets around the issue that larger farms are able to expand production while small ones cannot, it is useful to appeal to the notion of profit efficiency. Small farmers are most likely to be able to stay in business—and perhaps to gain market share—if they are more efficient users of farm resources, both in a technical sense (being on the production possibility frontier, given existing technology) and in an allocative sense (being at the right place on the production frontier, given prevailing prices). If small farms are more efficient users of farm resources to secure profits than are large farms, perhaps because they put more care into what they do or because they value their own time at less than the cost of hired labor, they have a cost advantage over large-scale producers that will be difficult to dislodge.

This process yields a measurable index of relative competitiveness: relative farm efficiency in securing profit. Ceteris paribus, farmers that are more efficient users of farm resources to secure profits are more likely to be able to maintain market share than are larger producers who are less efficient in the same sense. Over time, the more efficient are in a position to invest more into the farm enterprise and to grow, whatever their starting size.

A standard way of assessing farm-specific relative profit efficiency is to estimate a "profit frontier" across a sample of farms, then to measure how far each farm in the sample lies below the frontier. Conceptually, such a frontier can be thought of as a function mapping profit relative to input and output prices and quantities of nontraded factors of production, where each point is the maximum profit that a farm can achieve given those relative prices and access to resources. Given a set of prices, the average farm with that level of resources will fall below the frontier. Thus, an ordinary least squares regression on data from a sample of farms of different sizes of profits against input and output prices and fixed factors of production (land, labor, and so on) will always lie below the theoretical frontier. The frontier itself has to be estimated in some fashion, looking at data for farms that perform best at each level of resources. Fried, Lovell, and Schmidt (1993) described a variety of approaches.

The measurement of the "most efficient" farms can be improved by estimating a *stochastic* profit frontier, which allows for measurement error in the econometric estimation of the frontier itself and thus for the fact that observations for some farms will lie above the estimated "best" frontier (see Battese 1992; Coelli, Rao, and Battese 1998; and Kumbhakar and Lovell 2000 for surveys illustrating the development of this literature). In our case, the dependent variable is profit and the explanatory variables are farm-specific fixed resources (land, family labor, sunk capital), farm-specific input prices (feed, medicines, stock), and farm-specific output prices.

In the developing country situations studied, farm resources such as land may be nontradable inputs and must be accounted for in the frontier in terms of the amount available, not their price (because prices for these nontradable inputs do not vary much across farms). The unit prices received for output and prices paid for inputs can be expected to vary greatly, reflecting both quality differences and differential transaction costs facing different farm households. Examples of the latter are differences in the ability to search for, bargain for, and enforce sales contracts or to achieve market recognition from buyers through higher prices for the quality of their output when that quality cannot easily be observed at the time of sale, as is often the case for livestock products.

The actual performance of each farm in terms of profit can then be compared to an ideal performance level for that farm, given its resources and prevailing input and output prices. The difference between the ideal and the actual profit for that farm is the farm's relative profit.

Farm-specific profit efficiency (deviations below the frontier) are measured as the ratio of actual profit to the ideal (maximum) profit. The measure is bounded above by one (best, on the frontier) and less than one (zero, no profit). The upper bound can be achieved if a producer actually adopts a profit-maximizing combination of inputs and outputs, and the lower bound could be below zero because negative actual profit is possible (Kumbhakar and Lovell 2000).

If small farms have, on average, significantly higher profit efficiency when family labor is not costed, there is at least hope, if not reassurance for small farms. This is even truer if it holds when family labor is costed at the market wage rate. Stochastic profit frontier methodology allows us to go beyond simply making this determination; it also permits the investigation of which elements contribute most to explaining relative profit efficiency for large and small farms. Individual farms, large or small, may lie well below the profit frontier for reasons other than technical or allocative inefficiency. The profit frontier itself assumes that identical technologies are available to all farms. Technological change shifts the frontier up; if some farms in fact have a technology edge over others that is not accounted for in sample selection, it will shift the estimated frontier up for all farms, but the ones without that edge will be lower relative to the estimated frontier. Furthermore, farm-specific transaction cost barriers or policy distortions may also influence their position relative to the frontier.

A Methodology to Decompose the Determinants of Relative Profitability

Building on the Stochastic Profit Frontier Literature

This discussion leads us to the principal methodological approach, which is to estimate a stochastic profit frontier to derive a farm-specific measure of efficiency in securing profit, and then, following the lead of Jondrow et al. (1982), Ali and Flinn (1989), Battese and Coelli (1995), and Kumbhakar (2001), to identify factors that are associated with the profit efficiency of each farm, including the transaction cost barriers and farm-specific policy distortions faced by that farm.

The general form of the stochastic profit frontier used in this study to estimate profit efficiency is defined as

$$Y_i = f(X_i, W_i, P_i; \beta)\exp(v_i - u_i), \quad (1)$$

where Y_i = profit per farm i for a given commodity is defined as

$$Y_i = (P_iQ_i - C_iQ_i), \quad (2)$$

where P_iQ_i is total revenue from livestock activity per farm i in question (manure sales included); C_iQ_i is total variable costs, such as costs of feeds, fodder, day-old chicks (DOCs) or weanlings, hired labor, electricity, medicines, vaccines, water, depreciation, and so on, of securing revenue, excluding family labor per farm i; and Q_i is the quantity of output of the commodity in question per farm i; X_i = the vector of fixed factors used to obtain Y_i (for example, stock of family labor, land, buildings and equipment, and fixed capital stock, to control for differences in farm resources); W_i = the vector of farm-specific input prices; P_i = the weighted-average output price (the weights are the farm-specific transaction quantities);[1] β = the vector of unknown parameters to be estimated; and v_i, u_i are random error terms.[2]

The error term (v_i) is distributed independently and identically as a two-sided normal random variable around the frontier with mean zero and variance σ_v^2 to account for measurement error on both sides of the frontier, and inefficiency (u) is distributed independently as a one-sided (downward) random variable relative to the frontier to allow for the fact that farms in fact fall below the ideal efficiency. Average efficiency can easily be estimated by OLS regression of profit per farm against farm-specific input and output prices and farm-specific fixed factors of production. However, the frontier, showing ideal profits for any given level of farm resources and prevailing price level, can be estimated only if specific assumptions are made about the distributions of u and v across farms, and then only using a nonlinear estimation technique, such as maximum likelihood estimation (MLE). The critical assumption is the distribution of the u (inefficiency) term.

We adopt the approach of Battese and Coelli (1995), which allows for systematic differences across farms in the distribution of u, such that we do not assume away what we wish to investigate: viz., that transaction cost factors and policy distortions that are different for different farms help determine their relative profit inefficiency.

Battese and Coelli (1995) base their approach on the assumption that the expected value of the farm-specific inefficiency effect for farm i can be modeled as a function of farm-specific characteristics, which of course vary across farms, and fixed coefficients, which do not. In other words, $u_i \sim N$ (μ_i, σ_u^2), where $\mu_i = z_{ik}\delta_k$ is the mean of a truncated-normal distribution of u_i. The z_{ik} are k explanatory variables observed for farm i, associated with technical inefficiency effects (u_i), and δ is a vector of unknown coefficients to be estimated simultaneously with equation (1). Thus, the technical inefficiency effect, u_i in equation (1), can then be specified as $u_i = \delta_0 + \sum_{k=1}^{l} z_{ik}\delta_k + e_i$, where e_i is the inefficiency error term, defined by the truncation of the normal distribution with mean equal to zero and variance σ^2. The truncation of e_i occurs at $e_i \geq -z_{ik}\delta_k$ (Battese and Coelli 1995).[3]

Finally, the use of a stochastic profit frontier assumes that the underlying markets producing the observed input and output prices are competitive. This is justified in the case of markets for feed, stock, and output of hogs, chickens, and eggs in all four study countries. There are large numbers of producers and consumers of output and numerous physical markets and market outlets. On the input side, there are multiple feed millers/sellers (the principal input to monogastric livestock—at least six competing firms in the Philippines study zones, which of those in all four countries were the least likely to have a large number of choices available due to lower size and in-

---

[1]This procedure may introduce some endogeneity in the regressors, but any likely bias is judged to be minor compared to the error of ignoring major price differences faced by different farms for otherwise apparently similar inputs and outputs. A further issue is that price differences across farms may reflect a mix of genuine quality differences (different commodities) and buyer uncertainties about quality (transaction costs); there is no easy fix for this, and we shall return to the significance of this lack of separation in the discussion of results.

[2]Note that the frontiers are run separately by country and commodity. In most cases sample farms did not engage in producing for sale more than one of our chosen commodities. Where they did, possible economies of scope from joint production were not taken into account for practical reasons, although the model used could incorporate multiple outputs in a straightforward if somewhat laborious way (Coelli, Rao, and Battese 1998).

[3]The log-likelihood function of this model is presented in the appendix of Battese and Coelli (1993). Estimation of the likelihood function also requires the specification of a relationship among the variance parameters such that $\gamma = \text{var}(\mu_{it})/[\text{var}(vi) + \text{var}(\mu it)]$, where the parameter γ has value between 0 and 1.

comes). Feeder animal stocks are also available for a variety of sources, including directly among farmers in the case of piglets.

Dairy input markets (heifers and fodder) are fully competitive in the three dairy country cases studied (Brazil, India, and Thailand), as they are in most locations in the world. Farmers in these countries can also choose between several market outlets in disposing of raw milk, the form in which most milk is consumed in poorer areas. In the richer areas (such as Brazil), milk is necessarily sold to processors for pasteurization, but there is also more competition among multiple dairies in the richer areas than in the poorer ones. However, there is a built-in incentive in the milk business to sell through the same outlet in both glut (rainy) and lean seasons, because it is often harder to find buyers in the glut season in developing countries. Furthermore, milk sales are so subject to issues of market trust and reputation that spot sales among anonymous actors tend to be rare (Staal, Delgado, and Nicholson 1997). Nonetheless, despite these caveats, there is no systematic reason to suspect uncompetitive markets for inputs and outputs for milk, as in the case of monogastrics.

Specification of an Estimation Model

A translog profit frontier is used because of the flexibility it allows in estimating parameters where it is not desirable to build in through model specification rigid assumptions about substitution relationships among inputs and factors. As shown by Kumbhakar (2001), it is also important to preserve homogeneity in going from the underlying production relationships to the profit frontier, which the translog specification does. The full form of the model is

$$ln Y_i = ln\beta_o + \underset{\text{(dummies)}}{\sum_{l}^{L} \beta_i D_{il}} + \underset{\substack{\text{(output} \\ \text{prices)}}}{\beta_i ln P_i}$$

$$+ \underset{\text{(input prices)}}{\sum_{j=1}^{J} \beta_{ij} ln W_{ij}} + \underset{\text{(fixed factors)}}{\sum_{k=1}^{K} \theta_{ik} ln X_{ik}}$$

$$+ \underset{\text{(price-factor interactions)}}{\frac{1}{2}\sum_{k=1}^{K}\sum_{j=1}^{J}\alpha_{ikj} ln X_{ik} ln W_{ij}} \qquad (3)$$

$$+ \underset{\text{(factor interactions)}}{\frac{1}{2}\sum_{k=1}^{K}\varphi_{ik} ln X_{ik} ln X_{ik}}$$

$$+ \underset{\substack{\text{(random} \\ \text{error)}}}{v_i} - \underset{\substack{\text{(random technical} \\ \text{inefficiency effect)}}}{u_i,}$$

where Y_i is the profit of the ith farm defined in equation (2); P_i is the weighted output price for farm i for the commodity in question as defined earlier; W_{ij} is the price of input j (j = hired labor, capital, feeds, DOCs/weanlings, medicines, electricity, and other inputs used by the ith farm); X_{ik} is the fixed factor k used by the ith farm (k = the value of breeding stock, value of buildings and equipment, total farm labor in hours, land in hectares,[4] and other fixed capital stock); and D_{il} is a vector of dummy variables for farm i (l = production arrangement, gender, scale, region/zone, access to credit, access to information). The v_i, u_i are as previously defined. The α_{kj}, φ_{ik}, θ_{ik}, and β_{ij} are coefficients to be estimated by MLE using Frontier 4.1 software Model 2 (Coelli 1996).[5] Frontier 4.1 estimates all of the parameters in one step to overcome the inconsistency in the

---

[4]No effort was made to correct for differences in land quality; for monogastric production using purchased feed, this is not an issue, and for dairy production we did not have the information necessary to do this.

[5]STATA's command "frontier" with the option of a conditional mean model provides estimators for the parameters similar to that of Frontier 4.1 estimates.

assumptions about the distribution of the inefficiencies by defining the inefficiency effects as earlier, but they are then incorporated directly into the MLE. The predicted farm-specific efficiency estimate is retrieved from $E[exp(-u_i) | e_i]$.[6]

It can be noted, based on observations, that the technologies used by large and small farms in the sample appear on the surface to be similar in terms of animal genetics, purchased inputs such as feeds, and practices, and these technologies appear to be largely divisible (except perhaps for poultry). If there are, in fact, economies of scale in production, these are likely to be bound up with factors such as family labor or technological change that are not scale-neutral (Feder 1985). Family labor is explicitly included in the frontier.

In the case of hog production, we need to control for the fact that some farms do full-cycle operations (farrow-to-finish), while others do wean-to-finish, and still others do both. The total revenue and total cost figures used to compute total revenue allow for mixing different kinds of sales and costs for the same commodity (say, pigs), as do the average farm-specific output and input prices used on the right-hand side (RHS) of the stochastic profit frontier (SPF). In order to pool full-cycle and wean-to-finish producers, we use a binary dummy variable for the type of producer on the RHS of the SPF to allow for fixed differences in mean profits among activities on the same farm.

Translog and Cobb-Douglas profit frontiers share the use of logarithms in the dependent variables and thus do not handle cases of negative or zero profits. Yet it is not unreasonable to suppose—as turned out to be the case for some of the sampled farms—that some farms lose money in some years. In fact, there is no perfect fix for this problem, and we employ a lesser-of-the-evils approach that is adequate for our present purposes, which consists of adding the lowest constant scalar to all farms in a given country-commodity subsample that leads to positive profits for all farms in that subsample. The scalar is different for different subsamples, but constant within each country-commodity grouping.

The cases of negative average farm profits are few in most cases (fewer than 5 percent of the farms sampled in India, the Philippines, and Thailand; substantially more in Brazil, where the medium-sized layer farm, for example, just broke even), and they are proportionately not very negative relative to average farm profit (from 0.4 percent to 6.7 percent of mean profit below break-even for those farms reporting negative profits in these three countries, but higher in Brazil given that the sample average profit levels were near zero in the survey year). Thus, the resulting bias from this nonlinear transformation of the data in three of the four countries at least is judged to be of minor importance compared to the bias that would arise from using a less appropriate functional form or arbitrarily dropping the least efficient sample members.[7]

The technical inefficiency effects (u_i) generated in equation (3) are estimated within the MLE model specified above as

$$u_i = \delta_o + \delta_1 Z_1 + \delta_2 Z_2 \ldots + e_1, \qquad (4)$$

---

[6]See Coelli, Rao, and Battese (1998, 190), for a detailed discussion of procedures and issues. The mean farm efficiency for a specific subgroup, such as a quintile of the size distribution, is the arithmetic average of farm-specific measures within the group in question and is expressed as a percentage of the maximum (frontier) efficiency for farms of that resource level, given the prices faced.

[7]The problem for Brazil is more troublesome, but, faute de mieux, we have elected to remain consistent across the four countries. Conceptually, the impact on the Brazilian results will be to amplify the relative magnitude of differences in unit profits across farms, thus probably overstating the estimated numerical impact of regressors on relative profit efficiency.

where Z_i is the ith farm characteristic determining relative inefficiency and e_i is distributed as earlier.[8]

Whether variables belong in the first or second equation, equation (3) or equation (4) of the stochastic profit frontier model, is not clear-cut in all cases; the choice depends to some extent on where the line is drawn between resource endowment and access variables, and also where we wish to focus. The first equation defines the context within which farms perform and should include the price structure faced by all farms and their specific endowment of fixed farm resources. The latter are limited to land, family labor, and physical capital. These variables serve together to estimate both the frontier and where each farm falls relative to it.

As discussed earlier, some of these variables could be interpreted as representing the interaction between more than one factor, and we are not attempting to interpret the individual coefficients of equation (3). The role of the first equation of the model is both to derive a farm-specific measure of relative profit efficiency and to control for the usual price and endowment explanations of relative profit efficiency to allow a clear look in the inefficiency term, equation (4), at the access-related variables, such as experience, education, policy distortions, externalities, and so on.

It could be argued with some justification that some variables in equation (4), such as human capital, might as well have been included in equation (3). However, this would both cloud the interpretation of their individual contributions to explaining a given farm's divergence from the frontier and raise the issue of where to stop. If years of schooling of the household head are included in equation (3), why not the number

of training courses attended by the spouse and so forth?

The inefficiency term, equation (4), of the frontier as estimated here includes all the human capital and other variables that proxy interhousehold differences in access to assets and information. This allows the easier analysis of the separate effects of transaction costs (such as the inability to borrow because lenders cannot be reassured that they will be repaid) and externalities (such as the fact that small farms may do a better job of "paying" for the pollution they create than do larger ones) and policy distortions (such as whether large farms receive more subsidies than do small ones).

Thus, the RHS variables of the inefficiency term, that is, the Z, includes proxies of differential access to credit for capital/feeds, subsidized veterinary medicines, access to feed of known quality, subsidies and taxes (for differences in policy distortions), access to potential and other sources of income, access to markets for output, access to information, distance to nearest city or residential area, and a measure of the farm-specific internalization of negative environmental externalities produced by that farm (more on this later).

Three remaining methodological problems concern the measurement of the farm-specific data. First, some of the explanatory variables that we may wish to include in the inefficiency term may not be observable at all, or in any event very hard to observe. This is especially true of transaction cost and externality variables. Second, some of the explanatory variables that we may wish to include in the inefficiency term may be endogenous in the sense discussed earlier: the causality goes both ways, introducing simultaneity bias in estimation. This is par-

[8]We use the Battese and Coelli (1995) approach to the estimation of the efficiency effects (a function of u) as specified in the Frontier 4.1 Model 2 software downloaded from Coelli's webpage and discussed in Coelli, Rao, and Battese (1998). Kumbhakar (2001) introduces an elegant improvement to this approach that controls for possible endogeneity between technical and allocative efficiency effects. However, the computational effort to implement this was felt to exceed the likely gain in terms of the purpose of the present exercise.

ticularly a problem for environmental externalities, because farm-specific differences here will help to determine relative profit efficiency as we define it, but themselves may be a function of the latter in some cases. Third, contract farming may be quite prevalent in some cases, and this raises the issue of which output prices and which input prices to use, because these are accounting concepts rather than actual prices for contract farmers.

Measuring Farm-Specific Internalization of Environmental Externalities

Two problems arise in trying to account for the fact that some farms pollute more per unit of output than others do. The environmental externalities of livestock production are hard to measure, and in many cases are determined simultaneously with the level of actual profit per unit. An externality is defined here as a return to an economic agent where part of the cost (or benefit) of undertaking an activity accrues to another entity that is not compensated (or charged) in the market. Negative externalities may be created in the production process for animal agriculture through odor, flies, and the nutrient-loading effects on soil of manure that is either mishandled or supplied in excess. Producers capture the benefit of negative externalities by receiving payment for livestock output while not bearing the full costs of their enterprise in terms of the impact on surrounding communities of odor, flies, poor water quality, and so on. Producers who do not pay the full cost of production may show up as "more efficient" (in financial terms) than producers who are otherwise similar but internalize some of the externality by cleaning up after the enterprise or making compensatory payments to surrounding communities.

The first problem is how to measure the value to producers of not paying for pollution created, particularly if this differs by scale of farm, because it will lead to mis-

leading comparisons of unit profits across scale categories. Externalities of the sort involved are exceedingly difficult to measure. There is the difficulty of determining from which farm the pollution in the river came. There is the issue that farmers themselves suffer some of their own pollution, and this needs to be netted out of the externality. There is the issue that the negative effects of pollution carry over into future time periods. Physical measures of costs in terms of decreased sustainability are also very difficult. Furthermore, the true consequences for sustainability of a given amount of manure will differ by soil type, temperature, rainfall, and so forth.

In view of these many difficulties, it is not practical to attempt to measure actual negative externalities in the present study. Instead, we focus on differences across farms in the amount of externality "internalized" when a farmer invests in pollution abatement through handling manure and dead stock in an ecologically sound manner. Higher expenditure per unit of output on a given farm for abatement of environmental externalities, other things equal, should be inversely correlated with the incursion of net negative environmental externalities per unit of output under the assumptions stated earlier. Thus, a farm that spends more per unit of output on environmental abatement is postulated to incur fewer negative environmental externalities than a farm that spends less on environmental services per unit of output.

The heroic assumption that allows us to proxy environmental mitigation using the money value of manure management is that a given amount of manure of a given sort is equally polluting from whatever farm it comes, as long as it is not spread on fields (one's own or someone else's). This assumes that spreading manure on crops is uniformly good (despite run-off into watercourses in some cases) and ignores the fact that farms close to population centers and watercourses probably produce more ecological harm per ton of manure than those far from people and watercourses, other things equal. By the same logic, if we are willing to assume that the relationship is cardinal as well as ordinal—as in the statement that a US$1 per 100 kg of output in abatement on farm A is twice as environmentally friendly as US$0.50 per 100 kg of output on a different farm—we have a workable index that differentiates (inversely) across farms in the amount of negative environmental externalities incurred. The assumptions are not perfect, but the only feasible alternative, ignoring negative externalities in econometric production work altogether, seems worse.

The components that go into a measure of environmental mitigation include all costs of disposing of manure, such as the costs of water treatment, investment in lagoons, labor spent collecting and drying manure for sale (evaluated at market rates), rental of machinery used for manure disposal, regulations, taxes paid for abatement, and compliance in dealing with environmental problems.[9] In addition, the spreading of manure on crops is considered to transform a potential externality (pollution) into a positive contribution to soil structure and fertility. Costing this benefit is hard to do with accuracy.

The simple approach we have adopted is to value all manure sold for spreading on the fields of others (the reason it is purchased) at its sale value at the producing farmgate.

Manure spread on one's own fields is valued at what it could have been sold for at the farmgate.[10] Thus, if manure is spread in the field and has any market value (that is, people are not just dumping it), the latter is included in the internalization of the externality. The worst that any farm can do under this approach is to have no abatement expenditure at all, and this is in fact the case for many farms.

Having a working index of environmental mitigation creates a new problem and a new opportunity. The new problem is that this index, measured in local currency units, is in many cases determined simultaneously with profits. Thus, profit depends on environmental mitigation expenditures, but environmental mitigation expenditures are also influenced by profit. One approach to solving this problem is to create an instrumental variable for environmental mitigation by regressing it on a series of exogenous determinants of environmental mitigation, but the bivariate correlation between profits and mitigation was low (suggesting not too much of a problem), the proposed remedy was onerous, and this was eventually rejected as impractical for our present purposes.[11]

The measurement of environmental mitigation by the procedure described was motivated by the need to incorporate environmental factors in the empirical analysis of efficiency. However, more direct measures of environmental impact can be better used

---

[9]It might be thought that valuing family labor at market rates in this context is inconsistent with not valuing it in the efficiency regressions. The difference is in the purpose: for efficiency, we want to know if family farms can compete at any price. For valuing environmental mitigation, we need a cross-farm measure of the differential and largely optional efforts to clean up their mess. Both personal effort and hired labor are relevant expenditures here.

[10]Although spreading manure is counted as internalizing a negative externality, it may also boost own-farm profitability. It is not practical to net out benefits obtained from "expenditure" on mitigation, however, nor is it conceptually necessary, because the effort involved presumably still prevents pollution downstream, and presumably the person downstream does not care whether the farmer benefits as long as the pollution is stopped.

[11]Using instrumental variables in the present context is also not without problems, because it requires finding and testing for suitable instruments, finding procedures for handling zero expenditure cases, and dealing with the additional complexity in the error terms of the stochastic frontier. After time-consuming initial trials in the Thai case study that did not yield particular improvements, this approach was abandoned in favor of the much simpler and probably equally robust procedure of simply using the farm-specific estimate of mitigation expenditure on the RHS of the inefficiency term of the stochastic profit frontier model.

to look at the impact of livestock production on the environment and whether this differs across scales of farm. The next section explores a methodology for directly assessing the interaction of animal density and the environment.

Mass Balance Calculations

Theory of the Mass Balance Calculations Approach

The use and disposal of animal manure and dead animals has become a concern, as discussed in Chapter 2, as the structure of the industry has shifted toward fewer but larger operations, the percentage of animals raised in confinement has increased, and cities have expanded so that people are living closer to areas where livestock are raised. Traditionally, farmers applied manure to agricultural land to promote plant growth, thereby recycling much of the nutrients. With fewer but larger operations relying on inputs trucked in from the outside, the manure has become more concentrated in localized areas. When application rates exceed the carrying capacity of the land to assimilate nutrients, repeated applications can lead to a buildup of nutrients in the soil. This, in turn, increases the potential for nutrients to move from the field through leaching and runoff and to pollute groundwater (Kellogg et al. 2000).

To see whether a farmer has the ability to use all manure on his own farm, the farm's balance of manure nutrients relative to the farm's potential to use the nutrients through crop production is calculated based on the household surveys for each country. It is estimated that broilers and dairy produce 80 lbs (36 kg) of manure per day per 1,000 lbs (454 kg) liveweight animal unit, layers produce 60.5 lbs (27 kg) per day per animal unit, and swine produce 63.1 lbs (29 kg) per day per animal unit. This manure contains nitrogen, phosphorus, and potassium that, if not used or disposed of in a safe manner, can seep into the water table and cause groundwater pollution. Given the current livestock populations in the study countries and their expected growth, there may be a problem.

The amount of chemical fertilizer applied per land unit was also included, when available, and we computed the mass balance of nutrients applied to the land. Crop assimilation capacity was estimated to determine whether a crop could assimilate all the nutrients produced on-farm, and then the amount of manure sold off-farm, if any, was subtracted.

Animal Unit Calculations

Nutrient values from livestock were calculated based on animal units of 1,000 lbs (454 kg), the most commonly used metric to calculate nutrient levels in the United States. Animal unit conversions are used to equate excretion across species,[12] because the amount of manure produced by a cow is not equivalent in chemical terms or weight to the amount produced by a broiler or a hog. Table 3.1 illustrates that one U.S. animal unit is equivalent to one cow, five pigs, or 250 broilers or layers, a calculation based on averages used by the U.S. Agricultural Extension Service.[13]

---

[12]A U.S. animal unit is a unit of measure developed to compare differences in the amount of manure produced by species. One U.S. animal unit is generally defined as 1,000 pounds of live animal weight, but requirements for specific animals differ by species, age, weight, diet, and so on. For instance, while 250 chickens produce 298 pounds of nitrogen per year and 209 pounds of P_2O_5, it takes less than one cow to produce the equivalent (Pennsylvania State Cooperative Extension Service 1993).

[13]Different countries have different conversion factors and limits. For instance, the European Community Directive limits the number of manure-producing animals per hectare of land (for up to 16 weeks) to 2 dairy cows, 4 ground stock/beef cattle, 16 fattening pigs, 5 sows with piglets, 100 turkeys or ducks, 133 laying hens, or 285 ground hens. This is equivalent to a limit of 170 kg per ha per year of total nitrogen (including that deposited

Table 3.1   Converting animal liveweight to crop nutrients generated

Animal	Animals per animal unit (1,000 lbs. liveweight)	Amount of nitrogen generated per 1,000 lbs. liveweight per year (kg)	Amount of phosphorus generated per 1,000 lbs. liveweight per year (kg)
Layer	250	135.0	95.0
Broiler	250	135.0	95.0
Swine	5	68.0	54.0
Dairy cow	1	7.7	13.3

Source: Pennsylvania State Cooperative Extension Service (1993).

Total Nutrient Production

The total nutrient deposition from livestock for each household is estimated using these parameters, where the total nutrient deposited by household h is the sum of the nutrients produced by animal units of livestock type l belonging to household h. Data on commercial fertilizer use, if any, were added to the calculations to derive total nutrient use on-farm using the following formula:

$$T_h^n = \sum_l \alpha_l^n AU_{lh} + CF_h^n, \qquad (5)$$

where l = livestock category; n = nutrient type; h = household; T_h^n = total nutrient n deposited by household; AU_{lh} = animal units of livestock type l in household h; CF_h^n = form of nutrient n applied as commercial fertilizer by household h; and α_l^n = amount of nutrient n produced per animal unit of livestock type l.

Estimation of Crop Uptake

The capacity of household land to absorb nutrients is estimated assuming that all the available land is planted with a crop that would use the nutrients to determine whether a household would have the poten-

tial to use all the nutrients produced given its current number of animals.[14]

The capacity for each household to use the nutrients produced by its livestock is computed as the area or cropland available to the household multiplied by the nutrient uptake by the crops planted on the land. To determine this, we calculated the potential for either corn (for Brazil) or rice (for India, the Philippines, and Thailand) to take up these nutrients under the assumption that all the available cropland was planted with either corn (maize) or rice. It is estimated by this procedure that 140 kg of nitrogen and 56 kg of phosphorus would be removed per hectare of corn crop. It should be noted that the actual nitrogen requirements for corn production vary between 68 and 168 kg per hectare without irrigation and as much as 280 kg per hectare with irrigation. We assumed that the nitrogen uptake for rice production is 100 kg per hectare and the phosphorus uptake is 32 kg per hectare (Sims and Wolf 1994; Sims 1995).

Mass Balance

In order to determine the nutrient balance on the farm, the difference between manure nutrient production and consumption of the farm is calculated. The mass balance (MB)

---

while grazing) in zones deemed vulnerable to nitrate leaching (Williams 1995). Equivalent conversion factors might be lower for some developing countries, because the amount of nitrogen and phosphate excreted in animal manure depends on diet, species, and age of animal (Faasen and van Dijk 1987).

[14]This estimation inevitably overestimates what can be absorbed. Unfortunately, it is necessary to use this estimate because most of the survey data are not detailed enough to distinguish crop acreage from building area.

for each nutrient of interest (nitrogen and phosphorus) is expressed by the following equation:

$$MB_n = \beta^n A_h - M^n \times au_l \times N, \qquad (6)$$

where A_h = area of cropland owned or operated by household h; β^n = absorptive capacity for nutrient n per unit of land; M^n = nutrient n produced per animal unit; au_l = animal units for specific livestock l; and N = number of animals.

The result indicates a household's potential capacity to assimilate nutrients based on the current number of animals on their property. A positive mass balance implies that there is sufficient land to assimilate the nutrients produced, while a negative mass balance suggests that there is not enough land to absorb them. The results are also interpretable in physical terms. A negative mass balance of –10 metric tons for nitrogen, for example, implies that the farmer is loading nitrogen at the net rate of 10 metric tons per hectare per year, creating a surplus that will cause problems over time.

Manure Sold Off-Farm

Although manure is a potentially valuable fertilizer and soil conditioner, areas with concentrated livestock production may not have adequate cropland for nutrient use stemming from byproducts of livestock. In these cases, exporting nutrients from concentrated areas to surrounding areas may be both environmentally and economically beneficial. Markets for manure seem to exist in all countries, primarily for poultry manure used on horticultural crops, and to some degree for dry swine manure and dairy manure in some countries.

Animal waste tends to be bulky, difficult to apply, heterogeneous, and relatively poor in terms of nutrient content compared to the more homogeneous, higher nutrient content of commercial chemical fertilizers. This often makes it expensive to put animal waste into a form that can be transported, and the transportation costs tend to be high.

Further, because manure is often in liquid form (and tends to be sold in developing countries in dry form and is rarely treated), much of the nitrogen in it volatizes into the atmosphere prior to use.

The nitrogen in most fresh poultry manure is in the form of either ammonia or organic nitrogen. The concentration of nitrogen in poultry manure is the highest of all animal manures. This is due to the birds' common duct for urine and feces elimination and to the relatively low water content of the litter in broiler houses (Sims and Wolf 1994). Given that most broiler operations use bedding material in which the manure is dropped, it is relatively easy to bag and transport. Many poultry operators clean out facilities after each grow-out cycle, sometimes six times a year; thus the potential for facilitating the development of a market for poultry manure exists in many countries. Occasionally, poultry and swine manure is used directly as fish feed. Other uses of manure reported in the surveys were for flooring in buildings and for mushroom farming. There were limited markets for liquid swine manure, perhaps because there were considerable costs for disposing of bulky liquid manure.

The sample households, for the most part, did not collect dairy manure when animals were raised in pastures; in areas where they were not, such as India, it was collected and sold. Though there may be a market for manure, the market for unprocessed manure may be seasonal, because crops need fertilizer only in certain periods of their growing cycle during some times of the year.

Handling Contract Farming

Two complications arise from the inclusion of contract farmers in the empirical analysis. First, contract farmers do not pay for feed or other inputs, so their input prices are zero and their output prices are typically fees negotiated in advance. Thus, there are both conceptual and data problems in comparing independent and contract farmers in the same regression. The solution adopted for the price problem posed by contracting is to use actual prices paid for inputs (zero) and received for outputs (fees) for contract farmers. These balance each other in the computation of profit for the dependent variable. On the RHS of the first equation of the frontier model, equation (3), the zero input prices for contract farmers are handled using slope dummy variables for coefficients of the prices of inputs provided free to contract farmers, where the slope dummy applies if and only if the farmer in question is a contract farmer. This allows for the estimation within the same regression of different response coefficients to input prices for contract and independent farmers. Finally, particular attention has to be devoted to the fixed resources (land, family labor, sunk capital) provided by both contract and independent farmers. It is these variables that are central in explaining production levels of contract farmers, who are not constrained by price in expanding production.

The second problem is that the decision to be a contract farmer is presumably not independent from production decisions and profit, leading to possible simultaneity bias if a dummy variable for contract farming is included in the inefficiency term, equation (4), to explain why some farms are further from the frontier than others. The initial solution adopted for the endogeneity problem posed by contracting was to use an instrumented variable in place of a simple dummy variable.[15] However, as in the case of environmental mitigation and endogeneity, this econometric nicety introduced more problems than it solved, because the choice of adequate instruments is not straightforward and one source of possible bias in the error term of the stochastic frontier is replaced by another. Given the nature of the fix-up used for the first set of problems with contracting in the first equation of the frontier model, it was judged expedient to simply use the actual dummy for contracting in the inefficiency term of the frontier model.

[15]This was obtained by using a probit estimator on a regression of the decision to contract (one if a contractor, zero otherwise) on all exogenous household and farm characteristics available that might explain why a farmer chooses to be a contractor, including at least one variable relevant to the decision to be a contractor (such as nonfarm income) that is not one of the determinants of relative profitability. Then the predicted contracting status (contractor or independent) from the regression for each farm was used in lieu of a dummy variable based on the actual status of the farm.

CHAPTER 4

Data and Surveys

Primary and secondary data were used to answer the hypotheses posed in this study. The secondary data used in this study came from data published by government agencies, gray literature, and surveys reported in previous research and were used to support descriptions of the livestock sector given in the previous chapters. The primary data used came from farm household surveys conducted under the direction of the national team leaders in Brazil (Carmargo Barros), India (Mehta and Sharma), the Philippines (Costales), and Thailand (Poapongsakorn) in 2002/3. The data collected through the household-level farm surveys from each of these countries were used to conduct econometric analysis of factors affecting the scaling-up of livestock production, taking into account the impact of technological change and environmental externalities.

Sample Locations and Their Characteristics

The location of the farm household surveyed for each country was biased so that the regions surveyed in each country included a representative sample for each commodity being studied in order to capture the diversification of the scale of livestock operations for that specific commodity. In addition, the farms sampled represented the variety of production activities that took place at different stages of production for specific livestock commodities. For example, in the case of swine, the types of activities considered were growing piglets, raising fatteners, or a combination of both activities. In the case of layers, representative samples for farms growing chicks for hens, feeding hens for laying eggs, or a combination of both were well represented.

Sample Selection of Study Sites in Thailand

In Thailand, a cluster of six provinces located less than 250 kilometers from Bangkok were chosen representing the largest production density of broilers, eggs, and milk. These six provinces were Chachoengsao, Cholburi, and Srakaew in the east, Saraburi and Lopburi in the central area, and Korat in the northeast. At the time of the study, Chachoengsao was the largest producer of broilers and eggs, and it was the second-largest swine producer after Ratchaburi in the west. The provinces of Korat and Saraburi are the main milk-producing provinces. In recent years, the number of dairy farms has expanded in Korat (particularly in the eastern districts) and farther east to Srakaew and Lopburi. New farms for layers, broilers, and swine can now be found in the connecting provinces of Saraburi and Lopburi. Korat tends to be an old location for swine farms, but it is where a number of small-scale producers remain. Both old and new areas of production were included for comparison purposes to aid in the assessment of the impact of expansion of modern livestock production on the smallholders who use traditional methods. Contract farming is prevalent in Thai poultry production, and the observed forms of contractual arrangement were also considered ex post in the sample farms.

Sample Selection of Study Sites in the Philippines

In the Philippines, three regions were selected: Central Luzon, Southern Tagalog, and Northern Mindanao. The leading region for commercial hog production in the Philippines is Central Luzon, followed by Southern Tagalog. Four major hog-producing provinces in the Central Luzon region were surveyed: Nueva Ecija, Bulacan, Pampanga, and Tarlac. Three major hog producing provinces in Southern Tagalog were surveyed: Batangas, Laguna, and Rizal. Northern Mindanao, although not particularly dominant in terms of its livestock output volume, was included because the World Organization for Animal Health (OIE) recently certified Bukidnon, a main center of operation for integrators and large companies in the northern part of Mindanao, as free of FMD without vaccination.

Smallholder independent hog operations are still present in small numbers in these sample regions, but the level of commercial hog production activities is higher. Smallholder contracts were not found in Central Luzon, but were present in Southern Tagalog (Luzon), where few feed mills engaging in contract production with smallholders existed. Commercial-sized contract production arrangements exist, but there is great difficulty in obtaining production and marketing information. Large independent commercial hog production farms also exist in all three regions. The type of activities engaged in by hog producers was well represented ex post by the samples taken—namely, farrow-to-wean (piglet production), farrow-to-finish, and grow-to-finish activities or a combination of farrow-to-wean and farrow-to-finish operations (Costales et al. 2003).

Central Luzon also has the highest level broiler production in the country, followed by Southern Tagalog. Although commercial contract farming with large integrators dominates broiler production, there remain small-scale contracts with smaller integrators (with 6,000–10,000 birds). Both small-holder independent broiler production and independent commercial broiler operation are disappearing.

Sample Selection of Study Sites in India

The India dairy team surveyed the three leading milk-producing states of the country: Gujarat in the west region and Punjab and Haryana in the north region. These states are well-developed milk-producing regions in India, reflecting significant differences in organizational structure, with the traditional cooperative milk sheds in the west and newer milk sheds in the north. Gujarat is dominated by milk cooperatives that were strengthened under a dairy development program. Two of the highest-volume milk-producing districts in Gujarat, Mehsana and Kheda, were surveyed. Punjab and Haryana, on the other hand, are dominated by the private sector, and the presence of cooperatives is limited to a few areas. In the Punjab, two districts with potential for milk production growth, Ludhiana and Moga, were surveyed. The dairy cooperative sector has a strong presence in Ludhiana district, while milk from the milk-shed area in the Moga district largely goes to Nestlé (a private sector dairy plant). Although the dairy sector is well developed in Haryana, the cooperative movement is weak. The districts of Karnal and Jind were surveyed in Haryana to get a representative sample across this state.

The Indian poultry surveys were conducted in the states of Andhra Pradesh and Haryana. Poultry production has grown rapidly in these two states during the last three decades. These two states cover the national spectrum of poultry development and scales of activity, and both have considerable potential for future development. Andhra Pradesh ranks first among Indian states in egg and broiler production (5.8 billion eggs and 140 million broilers in 1997–98), while Haryana ranked tenth in egg production (637 million eggs in 1997–98) (Government of India 2000). The districts sampled from these states represent both high-concentration and low-concentration areas.

Sample Selection of Study Sites in Brazil

The Brazil team selected the southern, southeastern, and center-west regions to study. The southern region surveys were conducted in the states of Rio Grande de Sul, Santa Catarina, and Paraná. These states are the main production areas for meat and dairy products produced by both small- and medium-sized livestock farms. The southeastern region is considered the wealthiest and most industrialized region of the country. In this region, sample households producing milk, broilers, and eggs were selected in the states of São Paulo, Minas Gerais, Rio de Janeiro, and Espírito Santo. Most of the dairy samples came from large-scale farms located in Minas Gerais.

The center-west region surveys were conducted in the states of Mato Grosso, Mato Grosso do Sul, and Goiás. This is the region where both grain production and cattle ranching predominate, and it is considered to have considerable potential for the expansion of crops and animal production. Milk production in this region comes from animals producing both milk and beef, which were captured on the farms surveyed in Goias.

Commodities Selected

The commodities chosen for in-depth analysis in each country were broilers, eggs, swine, and milk. The choice of these commodities stemmed from the desire to stick with a manageable task but to observe the production of those items that had the following characteristics: (1) production was growing rapidly at the national level, (2) the commodities were important for the country concerned, (3) small-scale operators had traditionally and recently been involved in producing these commodities, and (4) there was at least anecdotal evidence that smallholders were being displaced from the sector in the sense that they were losing market share to larger operations or would soon do so. Based on these criteria, it was decided that swine producers would be surveyed in Brazil, the Philippines, and Thailand; broiler operations would be surveyed in all four countries; layer operations would be surveyed in Brazil, India, and Thailand, and milk producers would be surveyed in Brazil, India, and Thailand (see column 1 in Table 4.1).

Rationale for Size Category

Given the objective of assessing the outlook for smallholders in a rapidly increasing trend of livestock production, livestock farms were stratified ex ante into two broad size categories: large-scale and small-scale. The farm size classification was mainly based on the size distributions of farm livestock holdings for the country and commodity in question; the subjective judgment of the country teams also played a part, because comprehensive data on size distributions in the survey areas were usually not available. In general, small-scale farms were defined as those with holdings small enough that hired labor is not normally expected to be found. For instance, farms that are classified as large-scale in the Philippines would likely be considered relatively small-scale in Brazil. Scale-related stratification was finalized ex post to provide the best scale-related differentiation in each country as opposed to using a consistent definition of scale across countries. Further substratification was done ex post to distinguish contract from independent farms for countries and commodities where contract farming was observed.

The Thai team classified their sample farms into three to four categories: small, medium-low, medium-high, and large as defined in Table 4.1. For swine, small-scale farms had 1–100 head; medium-scale farms were divided into medium-low, with 101–500 head, and medium-high, with 501–1,000 head; and large-scale farms had more than 1,000 head. Layer farms operating with 10,000 birds or fewer were categorized as

Table 4.1   Size of operations by country and type of livestock farm, 2003

| Country/type of livestock farm | Small-scale (head) | Medium-scale (head) | | Large-scale/ commercial (head) |
		Medium-low	Medium-high	
Thailand				
Swine	1–100	101–500	501–1,000	More than 1,000
Broiler	1–5,000	5,001–10,000	10,001–20,000	More than 20,000
Dairy	1–20	21–30	31–50	More than 50
Layer	1–10,000	10,001–50,000		More than 50,000
Philippines				
Swine	100 or fewer	101–1,000		More than 1,000
Broiler	10,000 or fewer			More than 10,000
India				
Dairy	4 or fewer	5–10		More than 10
Broiler	Fewer than 10,000			10,000 or more
Layer	Fewer than 10,000			10,000 or more
Brazil				
Swine	100 or fewer	101–1,000		More than 1,000
Broiler	10,000 or fewer			More than 10,000
Layer	10,000 or fewer			More than 10,000
Dairy	50 or fewer	51–70		More than 70

Sources: Compiled from Camargo Barros et al. (2003); Costales et al. (2003); Mehta et al. (2003); Poapongsakorn et al. (2003); Sharma et al. (2003).

small in scale; those with 10,001–50,000 birds were sorted out as medium-scale; and those with more than 50,000 birds were classified as large-scale farms. Broiler farms were grouped into four farm sizes: small-scale, with 1–5,000 birds; medium-low, with 5,001–10,000 birds; medium-high, with 10,000–20,000 birds; and large-scale, with more than 20,000 birds. Dairy farms keeping 1–20 cows were labeled as small-scale farms; those keeping 21–50 cows were grouped as medium-scale farms, further dis-aggregated as medium-low (21–30 cows) and medium-high (31–50 cows); and farms with more than 50 cows were considered large-scale farms.

The India dairy team categorized their sample farms into three categories: small-scale (with 1 to 4 animals), medium-scale (with 4 to 10 animals), and large-scale (with more than 10 animals). For poultry farms in India, those with fewer than 10,000 birds

were categorized as small-scale farms, while those with more than 10,000 birds were categorized as large-scale.

The Brazil sample of swine farms was divided into small-scale farms, with 100 sows or fewer; medium-scale farms, with 101–1,000 sows; and large-scale operations, with more than 1,000 sows. For broilers and layers, small-scale operations were defined as those keeping an inventory of 10,000 birds or fewer, and those with more than 10,000 birds were classified as operating on a large scale. Small producers of milk were those with 50 cows or fewer, medium-scale farms were those with 51–70 cows, and large-scale farms were those keeping more than 70 cows.

In the Philippines, hog raisers operating with 100 sows or fewer were categorized as smallholders; beyond that, they were classified as large-scale/commercial raisers. Smallholder activities were defined as those that mainly use household resources (labor and land). The size of holdings was relatively small, so hired labor was not normally required or expected. The commercial sample, however, was disaggregated into two categories: medium-scale commercial raisers (101 to 1,000 head) and large-scale commercial (more than 1,000 head). Large-scale commercial raisers may differ from small-scale to medium commercial raisers in terms of the existence of economies of scale in feed mixing, breeding, and processing of output (integration to processing).

Broiler production in the Philippines has become largely commercial, with contract production the dominant arrangement. Minimum contract sizes have been set at about. Hence, farms that raised 10,000 birds or fewer were considered smallholders, while farms categorized as large-scale or commercial were those growing more than 10,000 birds.

Sample Size and Composition

A feasible sample size for each type of livestock farm (scale, species) was selected by balancing the number of farms of various populations in the survey areas, the design purpose of comparing statistically across scales, and the budget constraint for data collection of qualitative and quantitative information from each farm household surveyed. Table 4.2 shows the final size and composition of the household survey samples for each country after attrition. On average, close to 20 percent of the respondents in the Philippines and Thailand are replacements of the original sample households because of nonresponsiveness or hesitation to give away information related to profits and income. The samples were stratified according to scale of operation (small-scale, medium-scale, and large-scale or commercial) and type of production arrangement (independent and contract).

The India dairy team randomly sampled a total of 520 households consisting of 200 smallholders, 148 medium-scale farms, 108 large-scale farms, and 64 commercial or periurban dairy farms from Gujarat, Punjab, and Haryana. All sample farms operate through cooperatives.

The selection of a sample was quite difficult for the India poultry team, because information on the populations of poultry farms was limited. The team obtained information from different sources, such as the state directory of poultry farms in Haryana and Andhra Pradesh, the integrators' list, and the National Egg Coordination Council. To capture characteristics of poultry units, such as nearness to urban/rural areas and nearness to a main road, the team sampled 159 broiler farms, 110 of which were small-scale and 49 of which were large-scale. Of the 159 farm households, 136 were independents and 23 were under contract arrangement with the integrator. A total of 160 layer farms were sampled, consisting of 63 small-scale and 98 large-scale farms, all of which were operating independently.

For the Philippines, a total of 207 hog farms (110 smallholder farms and 97 large-scale/commercial farms) were surveyed from Central Luzon, Southern Tagalog, and Bukidnon. Of the 110 smallholders, 87 were independents and 23 contract farms; of the 97 large-scale/commercial farms, 67 were independents and 30 contract farms. Surveys also covered 116 broiler farms, constituting 62 smallholders and 54 large-scale farms. The sample broiler farms were further disaggregated as independent or contract farms, of which 54 were classified as independents and 62 as contract farms.

The Brazil broiler survey consisted of 235 farms, of which 201 were operating on a large scale, and only six farmers out of the total were producing as independents. The sample size of the egg producers (layers) was 89 distributed among states according to the methodology defined by the Brazilian Institute of Geography and Statistics (IBGE). Of the 89 sample farms, only 6 were small in scale, and 83 were large-scale farms. As

Table 4.2 Sampling distribution by country/region/province and type of livestock farm (number of farms)

Country/region/province	Swine				Broiler				Layer				Dairy				
	Small	Medium	Large	All	Small	Medium	Large	All	Small	Medium	Large	All	Small	Medium	Large	Commercial/periurban	All
Thailand	20	112	42	174	74	78	18	170	41	32	23	96	35	38	19		92
Eastern province	5	71	31	107	67	13	16	96	31	26	19	76	15	14			29
Central province	4	35	9	48	6	37	1	44	4	6	4	14	7	14	15		36
Northeastern province	11	6	2	19	1	28	1	30	6	0	0	6	13	10	4		27
Philippines	110		97	207	62		54	116									
Central Luzon	24		35	59	31		30	61									
Southern Tagalog	47		29	76	31		24	55									
Northern Mindanao	39		33	72													
India													200	148	108	64	520
North Zone													100	68	58	34	260
West Zone													100	80	50	30	260
India					110		49	159	63		98	161					
Andra Pradesh					71		9	80	22		58	80					
Haryana					39		40	79	41		40	81					
Brazil	1	31	161	193	34		201	235	6		83	89	64	31	65		160
South	1	29	73	103	27		90	117	3		17	20	46	9	18		73
Southeast		1	30	31	7		28	35	3		66	69	64	19	30		60
Center West		1	58	59	0		83	83	0		0	0	7	3	17		27

Sources: Compiled from Camargo Barros et al. (2003); Costales et al. 2003; Mehta et al. (2003); Poapongsakorn et al. (2003); Sharma et al. (2003).

for the swine survey, a sample of 193 farms was taken, stratified according to the number of parents, following the methodology defined by the IBGE. There were 161 large-scale farms sampled and 31 medium-scale farms. The sample farms were further divided into three groups: integrated under contract by companies (99 samples), integrated as members of cooperatives (37 samples), and independent (57 samples). The sample of dairy farmers included 160 producers, which were taken from selected states representing 76 percent of the milk production in Brazil. Small-scale dairy farmers are expressive in Rio Grande do Sul and Santa Catarina, while larger farms dominate in Paraná, São Paulo, Minas Gerais, and Goiás.

A total of 92 dairy farm households in Thailand were sampled, of which 35 were smallholders, 38 medium-scale farms, and 19 large-scale farms. The most common production arrangement in dairy production is contract farming with a cooperative. For swine-producing households, a total of 174 households were surveyed, of which 20 were classified as small-scale, 112 were medium-low-scale and medium-high-scale farms, and 42 were large-scale farms. Of the total sample, 125 farms were independent farms and the rest were contract farms divided into fee or wage-contract farms (30 samples) and forward-price-guaranteed farms (19 samples).

The sample for layer farms in Thailand was composed of 41 smallholders, 32 medium-scale farms, and 23 large-scale farms. Of these 96 farms, 88 were grouped as independents; the rest were subcontract farms. The broiler samples totaled 170, of which 74 were small-scale farms, 51 medium-low, 27 medium-high, and 18 large-scale farms. Almost the entire group of broiler farms (99 percent) were subcontractors (except for one independent and one integrator), either under forward-price contract or per-chick (fee) contract arrangements.

Types of Contracts for Poultry and Hogs Observed in the Study Countries

There are two main types of contracts for poultry and hogs that are proliferating in the study countries: fee (or wage) contracts (by animal or by weight) and forward-price contracts (guaranteed and/or with profit-sharing). They differ mainly in the mode of grower compensation, in the accounting for and shouldering of the growing stock and feeds, in the need to monitor production activities, and in the need for enforcement of actual deliveries. They also differ accordingly in incentives, penalties, risks, and provisions for defaults.

Fee or Wage Contracts

These contracts are mostly issued by the large multinational or national integrators; the scale of these contracts is generally approximately that of a "commercial" scale of operations (10,000 or more birds for broilers, 200 or more head of fatteners for hogs). However, there are fee contracts that cover as few as 6,000 birds in the Philippines and 4,000 in southern India.

In fee contracts, the integrator typically fully bears the costs of growing stock, feeds, and veterinary supplies and services. Thus, the prices of stock and feed are zero from the viewpoint of the grower, possibly leading to a temptation to resell them clandestinely or use inputs on private stock. The integrator bears all market risk and shares production risk with the grower. However, the grower typically does not share in the benefits of increasing output prices (or share in the losses due to falling output prices). An integrator needs to monitor production fairly closely to prevent avoidance of critical quality control interventions and to promote good animal husbandry practices by the grower and to avoid diversion by the grower of the integrator's inputs, such as feed, to noncontract uses.

The grower receives a guaranteed fixed fee for each live animal (in cash per bird or slaughtered hog or, in some cases, per kilogram liveweight) that is successfully harvested in a condition that is acceptable to the integrator for the purposes of live sale or slaughter. Under some contracts, payments by kilogram rather than per head are designed to give the grower a stake in feeding performance. To ensure effort by the grower, fee contracts also typically have built-in incentive and penalty clauses tied to the grower's ability to meet the integrator's set of specified minimum performance standards. These standards typically refer to feed conversion ratios, harvest recovery (percentage of live animals harvested), and average liveweight (for broilers) or average daily gain (for hogs). Compensation in addition to the fixed fee is given to the grower for meeting or surpassing each of the performance standards. For growers who fall below the set standards, corresponding amounts per bird or hog are subtracted from the fee.

While fee contracts may be attractive to growers, they have two disadvantages that limit their widespread use with smallholders. First, the onus on integrators to closely monitor production makes this an uninviting option for all but the most locally based integrators. Second, to be able to participate in fixed-fee contracts, a potential contract grower must typically post a bond per bird or head of animal with the integrator prior to engaging in the contract. The most common form of bond is a cash bond, verifiable as a deposit in a bank, or another financial instrument. The average cost of the bond per bird or head of hog is very close to the cost of one day-old-chick in broiler contracts and to the cost of one fattener (weanling) as delivered to the contract grower. If the grower defaults on the contract, the integrator keeps the bond.

Forward-Price and Profit-Sharing Contracts

In price contracts, while the integrator advances the cost of growing stock, feeds, and veterinary supplies and services, these are later charged in full to the contract grower at the time of harvest and sale of output, when all costs are accounted for, before compensation is paid. In essence, growing stock and feeds are provided by the integrator on credit. The stock used and feeds consumed are, in fact, evaluated at prevailing market prices, with a mark-up imposed for relevant charges (transport to the farm, cost of money for stock or feeds credit). Price contracts are more suitable when close supervision is not possible, because they reduce the incentives to divert integrator inputs to other uses. Four-fifths of Thailand's broiler contracts are now under a price guarantee system (Poapongsakorn et al. 2003). As in the case of wage contracts, market risk is born by the integrator. However, production risk (such as mortality or avian influenza culling) is now fully born by the grower. The integrator, however, now has to find ways to deal with the grower's incentive to prevent default when output market prices rise. In general, the integrator has the exclusive right to choose when and to whom to sell the harvest, as in fee contracts.

Solutions for getting around the problem of grower default when prices rise include a bonus for weight gain, as in Thailand, and profit sharing (50–50), as in the Philippines and India. Possibly greater ability to enforce contracts in Thailand and India may explain the popularity of price contracts in those countries. In the Philippines, price contracts are mostly undertaken by relatively small local feed millers with contract growers that they know well, with the scale of contracts generally approximately that of a "smallholder" scale of operations (in the Philippines for example, fewer than 10,000 birds for broilers, fewer than 100 head of fatteners for hogs).

One important difference for smallholders between price and fee contracts is that there typically are no prior bond requirements for engaging in price contracts, unlike in the case of fee contracts. The main deterrent to bad faith on the part of the grower in price contracts is that the costs of stock and feed are to be charged to the grower at the end of the cycle, whether the activity makes a profit or not. It is in the interest of both parties that the activity itself generate positive profit.

Timing and Extent of Survey

The household surveys were conducted in each country between October 2002 and January 2003. Questionnaires for each type of livestock producer (milk, swine, broiler, and egg) were pretested and administered to the household head or decisionmaker of the farm. The sets of information collected from the four countries were relatively similar (such as household characteristics, costs of production of livestock commodities [variable and fixed costs], sales of livestock commodities, farm management activities, and environmental management practices) for cross-country comparisons.

For Thailand, farm locations were obtained from the Department of Livestock Development and other local authorities, the milk and egg producer cooperatives, producer associations, university professors who served as consultants to farmers, and agribusiness companies (CP, Betagro, Panus Pokaphand, and Bangkok Feed Mill).

For India, the dairy survey variables included household demographic characteristics, land ownership, cropping patterns, agricultural production for *rabi* and *kharif* seasons in 2001–2,[1] livestock ownership, asset ownership, milk production and marketing, labor employment in dairying, feed and fodder use, animal health and breeding services, credit, and environment issues. Return visits to respondents were required to validate incomplete questionnaires or information, wherever possible. Relevant secondary information was also collected from published and unpublished sources, based on discussions with key stakeholders in the study area, to supplement the primary data collected from the sample households. Some local administration offices, State Milk Marketing Federations, and dairy plants were visited to obtain information to supplement the sample surveys.

The field survey for Indian poultry was carried out from October to December 2002. The data originating from the survey were checked and cross-examined for consistency, and wherever gaps or inconsistencies were noticed, return visits to concerned farm households were likewise made to fill these gaps.

For production patterns in the Philippines involving defined cycles, such as for broilers, the last or most recent production cycle was considered. For production patterns with no defined cycles (for example, overlapping batches of farrow-to-finish operations) for hogs, input-output performance was observed in the month of October 2002.

The Brazil team conducted their survey for broilers, hogs, milk, and egg-producing households from October 2002 to January 2003. The last production cycle was observed for defined production patterns, and for cases with no defined production cycles, the input-output performance in the month of October was also investigated (as in the Philippines).

---

[1]*Rabi* season crops are sown at the beginning of winter season (October–November) and harvested in March–April. *Kharif* season crops are sown at the beginning of the monsoon season in June–July and harvested at the end of monsoon season during September–October.

CHAPTER 5

A Profile of the Survey Samples

This chapter discusses findings from descriptive statistics of the farm household surveys for broiler, swine, egg, and milk production undertaken in Brazil, India, the Philippines, and Thailand that lend support to the various hypotheses laid out in Chapter 1. For instance, the descriptive analysis in this chapter allows for the investigation of Hypothesis 1: *Small-scale producers have higher or equal profits per unit of output than do large producers.* The descriptive statistics also provides a glimpse of apparent differences between small and large-scale farms and between independent and contract farms. The significance of these differences will be formally investigated in Chapter 8.

Scale-Related Differences in Prices of Inputs and Outputs across Scale of Farms

Differences in Input Prices

Differences in prices received and paid by farmers could indicate differences in transaction costs across farms. The inputs considered to compare differences in prices paid by farmers are feeds, growing stock, and labor. Feed costs comprise 65–70 percent of total production costs. Differences in unit price may be attributable to observed differences or believed differences in quality, differences in size or package, differences in market infrastructure, the ability to bargain for and enforce sales contracts or to achieve market recognition, and knowledge about the specific item being traded. The latter two are major contributors to the creation of transaction costs.

It can be argued that the difference in feed price across farm sizes could be due to differences in quality in addition to differences in the average quantity of individual purchases and the location of sources of feeds, but it is difficult to establish such hypotheses because the quality of feeds used by the sample farmers was not directly investigated in this study. In the case of milk production, medium- and large-scale farmers in Thailand paid considerably lower prices for concentrate feeds, an average of 4.6 baht per kg and 4.9 baht per kg, respectively, than small-scale farmers paid, which was 5.4 baht per kg, on average (Poapongsakorn et al. 2003). As for broilers, small-scale producers in Thailand paid a higher price for feeds, 9.1 baht per kg, than large-scale producers, 6.7 baht per kg (Poapongsakorn et al. 2003). Small-scale broiler growers in Brazil also paid a slightly higher price for feeds, 0.47 real per kg on average, than did large-scale growers, 0.46 real per kg (Camargo Barros et al. 2003). Similarly, in the case of swine, large-scale swine producers in the Philippines paid a lower weighted (weights are the farm-specific quantities of various feed used) average price for feeds, 13 pesos per kg, than small-scale producers paid, 17 pesos per kg (Costales et al. 2003).

In terms of feed efficiency, large-scale farms had a lower feed conversion ratio (FCR) than did small-scale farms. For example, in Brazil, large-scale broiler producers performed better in terms of feed use per kilogram of output, with an FCR of 1.88, compared to small-scale producers, with an FCR of 1.94 (Camargo Barros et al. 2003). This finding is not surprising because large-scale farms use more intensive technology and have better access to inputs than do small-scale farms; thus, the former have an advantage in minimizing production costs.

There are cases in which input prices per unit paid are not different by scale or by production arrangement. For example, in Brazil, the prices paid by swine farms, including contract farms, did not vary between small-scale and large-scale producers. In terms of feed efficiency, large-scale producers obtained a slightly higher FCR (2.3) than small-scale producers (2.2), but the mean difference is not statistically significant (with standard error [s.e.] = 0.09). Among swine producers, independents were less feed-efficient, with an FCR of 2.5, compared to cooperatives and contract farms, with FCRs of 2.3 and 2.2, respectively (Camargo Barros et al. 2003). This difference could be attributed to the rigidity of the contract farming system concerning feed quality, technical assistance, and genetic engineering.

Differences in concentrate feed prices across countries, such as broiler farm households in the Philippines that paid higher prices for feed, an average of 14.3 pesos per kg (US$0.28 per kg), than did those in Thailand, 9.1 baht per kg (US$0.21 per kg on average), or Brazil, 0.21 real per kg (US$0.16 per kg), may possibly reflect differences in import tariffs on corn (the in-quota tariff for corn is 35 percent in the Philippines and 20 percent in Thailand).

Another important input in livestock production is growing stock, such as day-old chicks (DOCs) and weanlings, which account for about 20 percent of the total production costs. In the case of the Philippines, smallholders paid relatively higher prices for DOCs (15.3 pesos per bird on average) than large-scale raisers, who could purchase the chicks for an average of 13.4 pesos each, the difference of which could be attributed to bulk purchasing (Costales et al. 2003). Interestingly, small contract growers paid much higher prices for DOCs compared to smallholders and large-scale growers, 16.9 pesos per bird, which was set by their small integrators. This may be because buyers believe such DOCs are more likely to be free from diseases. Another possible explanation is the margin added by small integrators who purchase DOCs from large integrators for resale. In Thailand, small-scale broiler producers paid a higher price for DOCs, 7.6 baht per chick, than did large-scale growers, 6.3 baht per chick (Poapongsakorn et al. 2003).

In the case of weanlings, hog contract growers in the Philippines, regardless of scale, paid higher prices for weanlings than did independent producers (Costales et al. 2003). On the other hand, the genetic characteristics of weanlings purchased by contract growers, and thus the predicted quality of output, were much more uniform and generally of higher quality from the standpoint of lower back fat. Comparing contract growers by farm size, larger-scale producers paid slightly less for weanlings than did small-scale contract growers.

As for other inputs, such as labor, small-scale producers are at an advantage because they employ mostly family labor. Camargo Barros et al. (2003) reported that large-scale broiler farms paid higher prices for inputs than did small-scale farms with hired labor. Large-scale farms paid higher wages, 652–742 reals per month, compared to small-scale farms, 400 reals per month. Large farms usually hire labor on a permanent basis for a year or for a cycle, while small farms hire on a casual, daily basis.

In most country and commodity cases, smallholders paid a higher price per unit of input, particularly for feeds and growing stock, than did large-scale farmers. Overall, large-scale farms (particularly contract farms) are more feed-efficient than small-scale ones because of better access to inputs (such as feeds of known quality), better management, and access to intensive and modern technology, particularly in terms of improved breeding stock.

Differences in Output Prices

As in the case of inputs, differences in output prices per unit often stem from actual differences in (observable) quality and the degree of the buyers' belief in that quality. Smallholder independents typically rely only on their buyers and other smallholders for price and market information, while contract farmers get such information from their integrators or cooperatives, aside from their peers. Dairy farmers in India, for example, sell a large proportion of their milk to cooperatives on a regular basis, with the assurance of timely payment. The cooperatives are then responsible for the sale and distribution of milk, normally to private buyers such as wholesalers and retailers.

Prices of Milk across Farms. In Brazil, large-scale dairy farmers received a higher price for milk, 0.4 real per liter, than did smallholders, 0.3 real per liter (Table 5.1). This seems to suggest that large-scale farmers have better bargaining power than do smallholders. Other possible reasons for this price difference are a better quality of milk as recognized by buyers and lower transport costs due to larger volumes transported by large-scale farmers compared to smallholders (Camargo Barros et al. 2003).

In Thailand, large-scale dairy farms received a slightly higher price, 12 baht per liter of milk, than the medium-sized farms, 11.4 baht per liter, and the small-scale farms, 11.7 baht per liter. The price variation can be attributed to differences in sale prices offered by the cooperatives. Although the output price is derived from the national guaranteed price of 13 baht per liter, it may vary depending on the services that cooperatives provide, such as free artificial insemination and veterinary services.

Buffalo milk in India commanded a higher price than cow's milk because of its high fat content (6 percent versus 4 percent, on average). The average price received by Indian dairy farm households for buffalo milk was about 11.6 rupees per liter in both regions, while the price received for cow's milk was 8.3 rupees per liter in the northern region and 7.5 rupees per liter in the western region. As shown in Table 5.1, price variation across farm sizes (small to commercial farms) can be observed in the northern region, but is not evident in the western region. This price impartiality is probably due to the strong presence and networking of dairy cooperatives in the western region, where cooperatives deal with small and large producers equally and procure milk at a price based on fat and solid-non-fat content, irrespective of the quantity of milk sold. In the northern region, the presence of an organized or cooperative sector is very limited, and producers depend on informal sector milk vendors, who purchase their output at the farmgate.

Prices of Live Broilers across Farms. In Thailand, the price of broilers per kilogram received by large-scale contract growers was higher than the price per kilogram received by small-scale contract growers (Table 5.2). In Brazil, large-scale producers received higher prices, 0.08 real per kg, than small-scale producers, 0.07 real per kg of output, possibly because of homogeneity of output and lower transport costs. But in the Philippines, smallholder independent producers received higher prices, 47.6 pesos per kg, compared to 44.7 pesos per kg received by large-scale independent producers (Table 5.2). Smallholder contract producers received slightly higher prices than commercial producers.

Prices of Hogs/Piglets across Farms. In Thailand, small-scale farms received lower prices for fattened hogs, 32 baht per kg liveweight, than did large-scale farms, 34 baht per kg liveweight, and the mean differ-

Table 5.1   Output price of milk across farm sizes by country, 2002

Variable	Farm size (number of cows)				
	Small (1–20 cows)	Medium (21–50 cows)	Large (more than 50 cows)		All
Thailand					
Output price (baht/liter)	11.7	11.4	12.0		11.6
Output price (US$/liter)	0.27	0.26	0.28		0.27
	Small (4 or fewer head)	Medium (5–10 head)	Large (more than 10 head)	Commercial[a] (more than 10 head)	
India					
North					
Buffalo milk (rupees/liter)	11.3	11.2	11.9	13.2	11.6
Buffalo milk (US$/liter)	0.23	0.23	0.24	0.27	0.24
Cow milk (rupees/liter)	7.5	8.3	8.6	8.6	8.3
Cow milk (US$/liter)	0.15	0.17	0.18	0.18	0.17
West					
Buffalo milk (rupees/liter)	11.3	11.2	11.6	12.9	11.6
Buffalo milk (US$/liter)	0.23	0.23	0.24	0.27	0.24
Cow milk (rupees/liter)	7.4	7.5	7.5	7.7	7.5
Cow milk (US$/liter)	0.15	0.15	0.15	0.16	0.15
	Small (50 or fewer head)	Large (more than 70 head)			
Brazil					
Output price (reals/liter)	0.3	0.4			
Output price (US$/liter)	0.10	0.14			

Sources: Compiled from Camargo Barros et al. (2003); Poapongsakorn et al. (2003); Sharma et al. (2003).

Notes: The currency conversion rates used are based on 2002 foreign exchange rates: for Thailand, US$1 = 42.96 baht; for India, US$1 = 48.61 rupees; for Brazil, US$1 = 2.92 reals.

[a]Commercial farms in the case of India are defined as those dairy farms that are close to the cities and have more than 10 head.

ence is statistically significant at 5 percent (with s.e. = 1.09) (Table 5.3). Interestingly, Costales et al. (2007), examining the Soro-soro Ibaba development cooperative in Southern Luzon (different from the current sample), found that large independent farms received higher liveweight prices for hogs than did smallholders, but that smallholder contract farmers selling on the open market through the feed mill integrator received the same prices as large farms. In the case of the current sample, smallholder contract farms received even higher prices (about 8 percent higher) than large farms.

In the case of piglets, large-scale farms received higher prices from sold piglets, 69 baht per kg liveweight, than did small-scale and medium-scale farms. Price-guarantee farms enjoyed higher prices of pigs and piglets than did independent farms (Poapongsakorn et al. 2003).

In the Philippines, the prices received for piglets sold by large-scale producers are higher, 97.5 to 112.4 pesos per kg, than the price received by independent smallholders, 92.5 pesos per kg (Table 5.3). Among independent producers, large-scale operators received higher prices, 112.4 pesos per kg, than do their medium-scale counterparts, but the difference is not statistically significant (with s.e. = 12.5). Except for the small contract growers, who received relatively

Table 5.2   Output price of live broilers across farm sizes by country, 2002

Variable	Farm size (number of birds)				
	Small (1–5,000 birds)	Medium (5,001–10,000 birds)	Large (10,001–20,000 birds)	Commercial (more than 20,000 birds)	All
Thailand					
Forward contract price (baht/kg)	25.3	25.7	26.0	26.3	25.6
Forward contract price (US$/kg)	0.6	0.6	0.6	0.6	0.6
Per-bird contract fee (baht/kg)	3.3	3.7	4.7	7.0	3.9
Per-bird contract fee (US$/kg)	0.08	0.09	0.11	0.16	0.09

	Small (10,000 or fewer birds)		Large (more than 10,000 birds)	
	Independent	Contract	Independent	Contract
Philippines				
From regular buyers (pesos/kg)	47.6		44.7	
From regular buyers (US$/kg)	0.9		0.9	
Integrators (pesos/kg)		6.3		6.1
Integrators (US$/kg)		0.12		0.12

	Small (10,000 or fewer birds)		Large (more than 10,000 birds)	
India				
From regular buyers (rupees/kg)	61.18		59.77	
Integrators (US$/kg)	1.26		1.23	
From regular buyers (rupees/kg)		3.80		4.38
Integrators (US$/kg)		0.08		0.09

	Small (10,000 or fewer birds)		Large (more than 10,000 birds)	
Brazil				
Contract fee (real/kg)		0.07		0.08
Contract fee (US$/kg)		0.02		0.03

Sources: Compiled from Camargo Barros et al. (2003); Costales et al. (2003); Poapongsakorn et al. (2003).

Note: The currency conversion rates used are based on 2002 foreign exchange rates: for Thailand, US$1 = 42.96 baht; for the Philippines, US$1 = 51.60 pesos; for India, US$1 = 48.61 rupees; for Brazil, US$1 = 2.92 reals.

higher prices, all producers received, on average, similar prices for fattened hogs, 52.2 to 52.6 pesos per kg.

In Brazil, small-scale and large-scale producers received the same price for fattened hogs, 1.1 real per kg liveweight, on average (Table 5.3). In some states such as in Mato Grosso do Sul (center-west), smallholders received a slightly higher price per kilogram liveweight of fattened hogs than did large-scale producers.

In sum, it was observed that large farms received higher prices for output per unit than did small farms, with the exception of independent broiler farms in India and the Philippines, where small farms have better offers than large ones. These price differentials can be attributed to different types of market outlets (including those organized by cooperatives and integrators), transaction costs associated with selling the output, and the quality of output as perceived by buyers.

Table 5.3 Output price of pigs and piglets across farm sizes by country, 2002

Type of animal	Farm size (number of head)			
	Small (100 or fewer head)	Medium-low (101–500 head)	Medium-high (501–1,000 head)	Large (more than 1,000 head)
Thailand				
Fattening (baht/kg)	31.8	34.4	34.5	34.1
Fattening (US$/kg)	0.7	0.8	0.8	0.8
Piglets (baht/kg)	66.4	54.9	60.4	69.1
Piglets (US$/kg)	1.6	1.3	1.4	1.6

	Small (100 or fewer head)		Large (more than 100 head)		
	Independent	Contract	Medium independent (101–1,000 head)	Large independent (more than 1,000 head)	Contract
Philippines					
Piglets (pesos/kg)	92.5		97.5	112.4	
Piglets (US$/kg)	1.8		1.9	2.2	
Slaughter hogs (pesos/kg)	52.6	56.6	52.4	52.2	52.5
Slaughter hogs (US$/kg)	1.0	1.1	1.0	1.0	1.0

	Small (100 or fewer head)	Large (more than 1,000 head)
Brazil		
Output price (reals/kg)	1.1	1.1
Output price (US$/kg)	0.38	0.38

Sources: Compiled from Camargo Barros et al. (2003); Costales et al. (2003); Poapongsakorn et al. (2003).
Note: The currency conversion rates used are based on 2002 foreign exchange rates: for Thailand, US$1 = 42.96 baht; for the Philippines, US$1 = 51.60 pesos; for Brazil, US$1 = 2.92 reals.

Scale-Related Differences in Access to Information, Markets, and Services

Differences in Access to Information

Differences across farm households in access to information and assets lead to differences in their cost structure for conducting exchange when buying inputs and selling outputs. We use the household head's age, education, experience, and participation in professional training related to livestock production as proxies to compare household differences in access to information.

The main findings of the four country studies showed that household heads of large-scale farms are older than household heads of small-scale farms (in most cases, except for large-scale dairy farmers in India and a few large-scale swine producers in Brazil, where smallholders are older than large-scale farmers). The data showed that heads of large-scale farm households who are older have eventually gained more years of experience in raising livestock compared to heads of small-scale farm households. It was also evident that heads of large-scale households typically attained higher levels of formal education compared to heads of small-scale households, except in the case of heads of small-scale broiler households in India.

On one hand, the low level of formal education of small-scale farmers could affect the cost of searching for information as well as the time spent to process and act on information received. On the other hand, years of schooling may be compensated with better farming skills, which could stem from work experience or from attendance in training courses such as new or improved dairy or poultry farming techniques. However, results from surveys showed that large-scale producers also tend to have more access to training than do small-scale producers.

In sum, it seems plausible that large-scale farm households bear lower transaction costs in buying inputs and selling outputs due to lower costs of searching for information. If so, it raises the question of whether, in the presence of higher transaction costs, smallholders will be able to compete with large-scale farmers, as will be investigated more formally in Chapter 8.

Differences in Remoteness of Farms by Scale of Operations

In this study, the general observation was that large-scale farms are usually located farther away from towns or communities and from public waterways than are small-scale farms. For example, in the case of Brazil, broiler and dairy farms are located as far as 31 km from the city. However, in this case location has no direct effect on prices of outputs and inputs. In fact, there were no price differences in the case of broilers, even if large-scale farms were located farther from town than small-scale farms. This result may be associated with the fact that most of the farms are contract farms, so contractors determine the price of output and supply feeds and medicines. Large- and small-scale farms were located at similar distances from towns and communities, and their prices of milk and feeds were relatively identical regardless of scale.

In the case of India, especially in the western region, most dairy farms are less than 1 km away from paved roads and have easy access to milk collection centers, due mainly to the good network of dairy cooperatives, which partly explains why the price of milk in this region is lower than in the north if we control for differences between cow and buffalo milk.

In the case of Thailand, sample smallholder broiler and dairy farms are located close to the nearest community (about 3 km) and about 1 km from public waterways. Even though small-scale farms are typically closer to towns, ceteris paribus, prices of inputs that they receive are higher than those received by large-scale farms (this is true for broiler farms and also for dairy farms but by a very small margin), which confirms the earlier assertion that price differentials between sizes of farms could be explained by bulk purchasing.

Differences in Access to Market Outlets

Livestock producers sell their products to different market outlets, such as cooperatives, processors, contractors, middlemen, wholesalers, retailers, traders/intermediaries, government extension services, and end consumers. In particular, independent farm producers sell most of their livestock products to private buyers such as wholesalers (merchants, supermarkets), retailers (vendors), and restaurants and fast-food chains.

In choosing a market outlet, livestock producers take into consideration factors that they perceive to be important, such as regularity in procuring sales, reasonable price offers, and sensible terms of payment. Prices of output received by producers could be lower or higher than the prevailing market prices, depending on their bargaining power, the type of market outlet, and, to some extent, the quality of their output for sale.

The sample small-scale farms were located closer to market outlets than were the sample large-scale farms. In the Philippines, for example, small-scale and small independent broiler and hog farms sell their output at a market 4–15 km away from the production site. Contract and independent large-scale broiler farms, on the other hand, have a main output market more than 30 km away from the closest town, on average. Large independent growers mostly rely on wholesalers/traders who also pick up the output; hence, they receive a lower output price (particularly for broilers) compared to independent smallholders, whose regular buyers are meat dealers/retailers located in nearby towns.

In Brazil, market outlets for slaughter hogs depend on animal weight. For instance, for lightweight pigs (<110 kg liveweight), the common market outlet is the fresh meat market shop. For heavyweight pigs (>110 kg liveweight), the market outlet is the meat-processing plant. Some large-scale producers, such as the large-scale egg producers in Brazil, sell their output directly to supermarkets, which give them a better deal on price (Camargo Barros et al. 2003).

In the case of India, especially in the western region, most dairy farms are less than 1 km away from paved roads and have easy access to milk collection centers, due mainly to the good network of dairy cooperatives, which partly explains why the price of cow's milk in this region is lower than in the north. Being near market outlets is an advantage, especially for perishable products such as milk. There is also a high proportion of dairy farmers in India who sell milk to unorganized sectors such as private buyers, subcontractors, and local vendors. Private buyers such as these are likewise present in Thailand. They buy milk at the guaranteed price just as the cooperatives do, but do not provide other services that cooperatives provide, such as the ones discussed in the following subsection.

Differential Access to Credit and Veterinary Services

Large-scale and small-scale farms, both contract and independent, clearly have distinct and differential access to credit and veterinary services. Large-scale contract growers have automatic access to credit for capital and breeding stock and for variable inputs such as DOCs and weanlings, feeds, and medicines because the integrators usually finance or supply these inputs. Although not all integrators can provide credit for capital and breeding stock to their contract

growers, capital investments can be financed through bank loans. In addition, large-scale farmers can easily gain approval for loans in public or private banks. One possible reason is that large-scale farmers own bigger landholdings than smallholders that could be used as physical collateral for formal loans. This is supported in the Philippine case, where data show that a higher proportion of large-scale farms take out formal sector loans than is the case for small farms, as might be expected (Costales et al. 2003). Moreover, the rigid mechanism for obtaining loans, such as establishing a credit line and repayment scheme and the high interest rates, are not compatible with the capacity and resources of small-scale farmers.

For all sample farms, whether large- or small-scale, the most common sources of credit are cooperatives, integrators, public banks, private banks, and relatives. On the other hand, smallholders, specifically the independents, are at a disadvantage in terms of access to credit and other production services because they have difficulty in establishing a credit history or providing other proof of creditworthiness.

Contract growers enjoy better access to other production support services than do independent growers. Veterinary services are an example; veterinary services are typically mandatory for contract farms and are supplied by integrators as stipulated in the contract agreements; independents, on the other hand, have to pay on a fee-for-service basis to avail themselves of these services.

In the case of India, the central government subsidizes livestock services for dairy farms. There are government veterinary hospitals/polyclinics, mobile and stationary veterinary dispensaries, and veterinary first aid centers in the country that provide animal health care and breeding facilities. These services are supplemented by those specific to cooperatives, private sector firms, and nongovernment organizations.

From the findings reported earlier, it can be deduced that small-scale producers are characterized by lower levels of education, fewer years of experience in livestock production, more moderate use of inputs per unit of output, more limited access to input and output markets, higher input prices, lower output prices, relatively smaller landholdings, and more limited access to credit for capital than are large-scale producers. Will smallholders be able to maintain profits per unit of output comparable to those received by large-scale producers under these conditions, and if so, why? The following discussion provides insights that help us provide an answer to this intriguing question.

Comparative Profit Performance

This section directly investigates Hypothesis 1, allowing us to determine whether small-scale producers have higher or equal profits per unit of output than do large-scale producers. This hypothesis is investigated using calculations of average profits per unit of output by scale category for each type of livestock commodity operation (dairy products, swine, layers, and broilers) and then by contractors versus independents. Profit (gross revenue per unit less variable cost per unit), as defined in Chapter 3, was computed directly from the farm survey data for each farm and then averaged, and the result was used in the financial (nominal) sense rather than with regard to social prices. Profit was calculated without costing family labor. As explained in Chapter 3, if small farms can produce at lower cost per unit than large farms by not costing family labor, there is at least a chance for them to stay in the market. For comparison,

an imputed cost of family labor (the prevailing agricultural market wage rate) was used selectively for cases in which suitable data were available to investigate the difference in results when family labor is and is not costed.

Milk

In the case of milk production, small-scale dairy farms in India had higher profits on average, 2.45 rupees per liter, without costing family labor, than large-scale farms, 0.52 rupees per liter (Table 5.4). Similar results at a lower magnitude held if family labor was costed at market rates. In Thailand, medium-scale dairy farms made about 20 percent more profit, 6.25 baht per liter, than either small- or large-scale dairy farms (Table 5.4). This profit difference can partly be explained by farmers' ability to use the right mixture of concentrate feeds and roughage. Medium-sized farms, for instance, gained the highest profit per kilogram of milk because they had the lowest feed cost per kilogram.[1] In Brazil, the profit per unit of output between small and medium dairy farms was the same on average, 0.04 reals per liter, while large dairy farms earned 0.05 reals per liter of positive profit (Table 5.4).

Among the countries listed in Table 5.4, dairy farmers from Thailand gained a much higher average unit profit in U.S. dollars per liter of milk at all scales of production than did those from India and Brazil. This is probably because of the much higher level of protection of dairy farmers in Thailand. Thai dairy farmers who contract with cooperatives (88 percent of the total sample) are fairly independent from the cooperatives in managing their farms. The role of their cooperatives is to provide them with services

---

[1]When feed was produced by the farm household, the price of feed was estimated by calculating the weighted average of market prices paid by the farmers for various types of concentrates and other feed supplements fed to the animal (weights are the farm-specific quantities of feed used). Likewise, the price of fodder was derived by taking the weighted average of market prices paid by the farmers for different kinds of green fodder and dry fodder fed to cows and/or buffaloes (weights are the farm-specific quantities of fodder used).

Table 5.4   Average profit per liter of milk across farm sizes by country, 2002

Profit	Region	Small (fewer than 10 liters)	Medium (10–20 liters)	Medium (21–40 liters)	Large/commercial (41–80 liters)	Large/commercial (81–150 liters)	Large/commercial (more than 150 liters)	All
India								
Average profit without family labor cost	North (rupees/liter)	2.21	1.53	1.10	0.88	0.63	0.38	1.47
	West (rupees/liter)	3.09	1.72	1.09	0.71	0.48	0.33	1.27
	Pooled (rupees/liter)	2.45	1.62	1.09	0.71	0.48	0.38	1.37
	Pooled (US$/liter)	(0.05)	(0.03)	(0.02)	(0.01)	(0.01)	(0.01)	(0.03)
Average profit with family labor cost	North (rupees/liter)	0.46	0.50	0.49	0.39	0.13	0.25	0.44
	West (rupees/liter)	1.45	0.37	0.69	0.62	0.47	0.38	0.66
	Pooled (rupees/liter)	0.52	0.42	0.53	0.49	0.40	0.29	0.43

	Farm size (number of head)			
	(1–20 head)	(21–50 head)	(more than 50 head)	All
Thailand				
Average profit (baht/liter)	5.10	6.25	5.35	5.63
Average profit (US$/liter)	(0.12)	(0.15)	(0.12)	(0.13)

	(50 or fewer head)	(51–70 head)	(more than 70 head)
Brazil			
Average profit (reals/liter)	0.04	0.04	0.05
Average profit (US$/liter)	(0.01)	(0.02)	(0.02)

Sources: Compiled from Camargo Barros et al. (2003); Poapongsakorn et al. (2003); Sharma et al. (2003).

Notes: Numbers in parentheses are average profit in US$ per liter. The currency conversion rates used are based on 2002 foreign exchange rates: for Thailand, US$1 = 42.96 baht; for India, US$1 = 48.61 rupees; for Brazil, US$1 = 2.92 reals.

such as loans, technical consultation, and buyers who could offer guaranteed prices.

Broilers and Eggs

As in the case of broilers, independent small-scale producers in India made more profits per unit than did large-scale independents (without costing family labor), 13.1 versus 10.9 rupees per bird, respectively (Table 5.5). But small-scale contractors did much worse, 1 rupee per bird, compare to large-scale contractors, 3.2 rupees per bird (Table 5.5).

Similarly, in the Philippines, independent smallholders also had significantly (at the 5 percent level of significance, with s.e. = 1.5) higher profits per kilogram than did large-scale independents, 1.6 pesos per kg versus 1.1 pesos (Table 5.5). Contrary to the situation in India, small contract broiler farms in the Philippines had higher profits per unit than did large contract farms, 4.1 pesos versus 4.0 pesos per bird, respectively (Table 5.5), and it is not surprising that the mean difference between unit profits of small and large contractors was not significant (with s.e. = 0.68).

In Thailand, large independent broiler farms made higher profits, 2.5 baht per kg on average, than medium-sized independent farms, 1.6 baht per kg (Table 5.5). If family labor is not costed, fee-contract farmers in the Thai broiler farm sample had similar profits per unit for large- and small-scale farms, 1.6 versus 1.5 baht per kg, respectively, but the medium-scale contractors had

Table 5.5   Average profit per unit of output of liveweight broilers across farm sizes by country and production arrangement, 2002

Profit	Unit	Farm size (number of birds)			
		Small (fewer than 10,000 birds)		Large/commercial (10,000 or more birds)	
		Independent	Contract	Independent	Contract
India					
Average profit without	(rupees/bird)	13.13	1.03	10.93	3.16
family labor cost	(US$/kg)[a]	(0.11)	(0.01)	(0.09)	(0.03)
	(rupees/bird)	11.36		9.98	
	(US$/kg)[a]	(0.10)		(0.09)	
Average profit with	(rupees/bird)	12.40	0.04	10.80	3.01
family labor cost	(US$/kg)	(0.11)	(0.003)	(0.09)	(0.03)
	(rupees/bird)	10.59		9.85	
	(US$/kg)	(0.09)		(0.08)	

		Farm size (number of birds)			
		Small (10,000 or fewer birds)		Large (more than 10,000 birds)	
		Independent	Contract	Independent	Contract
Philippines					
Average profit without	(pesos/kg)	1.59	4.05	1.07	3.96
family labor cost	(US$/kg)	(0.03)	(0.08)	(0.02)	(0.08)
Average profit with	(pesos/kg)	1.34	3.98	1.06	3.95
family labor cost	(US$/kg)	(0.03)	(0.08)	(0.02)	(0.08)
Brazil					
Average profit	(reals/kg liveweight)		0.05		0.06
	(US$/kg liveweight)		(0.02)		(0.02)

		Small/medium-low		Medium-high/large	
		Forward contract and independent	Per-bird wage contract	Forward contract and independent	Per-bird wage contract
Thailand					
Average profit	(baht/kg liveweight)	0.71	1.35	2.48	1.51
	(US$/kg liveweight)	(0.02)	(0.03)	(0.06)	(0.04)

Sources: Compiled from Camargo Barros et al. (2003); Costales et al. (2003); Mehta et al. (2003); Poapongsakorn et al. (2003).

Notes: Numbers in parentheses are average profit in US$ per unit of output. The currency conversion rates used are based on 2002 foreign exchange rates: for Thailand, US$1 = 42.96 baht; for India, US$1 = 48.61 rupees; for Brazil, US$1 = 2.92 reals; for the Philippines, US$1 = 51.60 pesos.

[a]Assuming one bird weighs 2.4 kg liveweight.

only 1.1 baht per kg profit (Table 5.5). The medium-scale farmers had to employ labor, which cut their unit profits relative to small-holders, but they were not in the same cost structure as the larger farmers.

In Brazil, small and large broiler farms had surprisingly similar average profits per unit, 0.05 real per kg versus 0.06 real (Table 5.5); this may reflect the facts that almost all broiler production in Brazil is vertically integrated, with the integrators supplying all the main inputs into production, giving producers access to better inputs and modern technology, and that integrators passed on some of their cost savings in dealing with larger producers to them. Smallholders maintained their unit profits close to the rates of large farmers by not costing family labor.

In Thailand, smallholder layer farms had higher profits per egg than did large-scale farms, although the mean difference was not statistically significant (with s.e. = 0.03) (Table 5.6). Among Indian layer

farmers, smallholders had higher profits per unit than larger farms if family labor is not costed, 0.23 rupees per egg versus 0.17 rupees (Table 5.6). If family labor is costed at market rates, smallholders had net losses per egg, while large-scale farmers continued to have positive profits. In Brazil, both large and small layer farms had net losses in the survey year, although the large farms almost covered their costs at a net loss of 0.01 real per egg, compared to a net loss for smallholders of 0.04 real per egg, not costing family labor (Table 5.6). One possible explanation for this outcome is the devaluation of the Brazilian real, which led to increases in prices of inputs that eventually affected farm profits (Camargo Barros et al. 2003).

Swine

It should be noted that profits were computed on the basis of what farmers had reported at the time of the survey in 2002/3,

Table 5.6 Average profit per egg across farm sizes by country, 2002

Profit	Unit	Farm size (number of birds)	
		Small **(fewer than 10,000 birds)**	**Large** **(10,000 or more birds)**
India			
Average profit without family labor cost	(rupees per egg)	0.23	0.17
	(US$ per egg)	(0.005)	(0.003)
Average profit with family labor cost	(rupees per egg)	−0.07	0.03
	(US$ per egg)	(−0.001)	(0.001)
		(10,000 or fewer birds)	**(more than 50,000 birds)**
Thailand			
Average profit	(baht per egg)	0.22	0.21
	(US$ per egg)	(0.005)	(0.005)
		(10,000 or fewer birds)	**(more than 10,000 birds)**
Brazil			
Average profit	(reals per box)[a]	−1.32	−0.33
	(US$ per egg)	(−0.013)	(−0.003)

Sources: Compiled from Camargo Barros et al. (2003); Mehta et al. (2003); Poapongsakorn et al. (2003).
Notes: Numbers in parentheses are average profit in US$ per unit of output. The currency conversion rates used are based on 2002 foreign exchange rates: for India, US$1 = 48.61 rupees; for Thailand, US$1 = 42.96 baht; for Brazil, US$1 = 2.92 reals.
[a]One box contains 360 eggs.

when some farms experienced negative profits in a very tough price and cost environment for poultry and swine producers, at world prices, and contracting probably looked relatively good from the farmers' perspective. For example, in the case of Brazil, large-scale swine farms (independents and contractors) lost an average of 0.15 real per kg of swine, while smallholders lost an average of 0.25 real per kg of swine (Table 5.7). Camargo Barros et al. (2003) pointed out that the profit per kilogram of output was higher for integrated/cooperative producers, 0.04 real per kg of output, than for independent producers, –0.40 real per kg of output. In fact, it was the integrated farms that actually made positive profits, while independent farms and farms under cooperatives incurred losses, –0.20 real per kg of output for the latter.

In the Philippines, independent smallholder swine farmers had higher profits per kilogram than large-scale independents, 26.6 pesos per kg liveweight versus 19.8 pesos (Table 5.7). Imputing the opportunity cost of family labor shows that large-scale farms had essentially similar profit rates per kilogram vis-à-vis when family labor was not costed. Adjusting for the cost of family labor affected smallholder farms' profit per unit of output much more than that of large-scale farms, as was expected, because the value of family labor input is by definition less the greater the output of the family.

Table 5.7   Profit per kilogram liveweight of output of swine across farm sizes by country and production arrangement, 2002

		Farm size (number of head)				
		Small (100 or fewer head)		Medium/large (more than 100 head)		
				Medium independent (101–1,000 head)	Large independent (more than 1,000 head)	Contract
Profit	Unit	Independent	Contract			
Philippines						
Average profit without family labor cost	(pesos/kg)	26.60	2.08	19.61	19.83	2.33
	(US$/kg)	(0.52)	(0.04)	(0.38)	(0.38)	(0.05)
Average profit with family labor cost	(pesos/kg)	26.45	2.05	19.58	19.82	2.33
	(US$/kg)	(0.51)	(0.04)	(0.38)	(0.38)	(0.05)

| | | Small/medium-low (500 or fewer head) | | Medium-high/large (more than 500 head) | |
		Independent	Contract	Independent	Contract
Thailand					
Average profit	(baht/kg)	11.9	11.5	15.4	1.7
	(US$/kg)	(0.28)	(0.27)	(0.36)	(0.04)

		Small/medium (1,000 or fewer head)	Large (more than 1,000 head)
Brazil			
Average profit	(reals/kg)	–0.25	–0.15
	(US$/kg)	(–0.09)	(–0.05)

Sources: Compiled from Camargo Barros et al. (2003); Costales et al. (2003); Poapongsakorn et al. (2003).
Notes: Numbers in parentheses are average profit in US$ per unit of output. The currency conversion rates used are based on 2002 foreign exchange rates: for Thailand, US$1 = 42.96 baht; for Brazil, US$1 = 2.92 reals; for the Philippines, US$1 = 51.60 pesos.

In Thailand, small-scale independent swine farms had average profits of 11.5 baht per kg compared to 20.0 baht for medium-sized farms and 15.4 baht for the largest farms (Table 5.7). On the other hand, large contract swine producers in Thailand generated the lowest farm profits per kilogram of output. A relatively large scale of operations compensated small margins per unit of output.

Comparative Profit Performance

From the findings just reviewed, it can be concluded that smallholders typically have higher profits per unit of output than do large-scale producers. This result was seen in the Philippines and India cases, where independent smallholders did better in terms of profits per kilogram of output than did large-scale farmers; also, profits per unit made by smallholder contract farmers are higher than those made by large-scale contract farms in the case of swine in Thailand and broilers in the Philippines. Therefore, smallholders will at least have a chance to compete with larger-scale producers because they have the ability to produce at a lower cost per unit of production or at least to achieve profits per unit of output similar to those of large-scale farmers, with or without costing family labor. This is a necessary condition for smallholders to survive, because it means that large-scale farms cannot easily drive smallholders out of the market by driving their returns below levels that they can live on, given their small production volume. However, it is not a sufficient condition for smallholders' survival, because if large-scale producers are more efficient on average, they will be able to eventually drive their costs down and survive on smaller unit profits but larger volumes of sales. Thus, the key issue is whether smallholders have the ability to use their farm resources more efficiently than large-scale producers. This comparison of relative profit efficiency is discussed in the econometric estimates of Chapter 7.

CHAPTER 6

The Impact of Scaling-Up of Livestock Production on the Environment

A major problem associated with increased livestock production is that it also results in an increased amount of manure and dead animals during the grow-out period, which may result in environmental problems if these byproducts of the production process are improperly used or disposed of.

As a fertilizer, manure is a valuable source of nitrogen, phosphorus, and organic material, and it improves soil structure and water retention on leached tropical soils. It can be a source of positive externalities in some cases. However, too much of a good thing can have a negative effect as the amount of manure applied increases. Furthermore, some soils benefit less than others. Nutrients from animal byproducts are harmful when the excess nitrogen and/ or phosphorus in them is not taken up by plants and leaches into the groundwater or contaminates surface water, and it can also lead to bacterial contamination of surface water and groundwater.

Water pollution may occur if nutrients from manure enter the water table because they are improperly applied to fields or disposed of, or it may be associated with improper disposition of dead animals that consequently release nutrients into the groundwater as they decompose. Air pollution may result as the nitrogen in manure is converted to ammonium or through the incineration of dead animals. Land degradation may occur if the carrying capacity of animals is too high, leading to overgrazing in the case of ruminants, or if the application of nutrients over a number of years causes buildup of nitrogen, phosphorus, and salt, resulting in reduced crop yields.

Disposal of dead animals is also a common environmental issue for monogastric livestock production. Normal mortality for broilers is 5 percent over the production cycle; under extreme conditions, such as great heat or an epidemic, it can be as high as 25 percent a day. Mortality rates per cycle for sows under developing country conditions are 5 percent, for preweaning piglets 19 percent, for layers 8 percent, and for dairy cows 4 percent. In the countries studied, these dead animals are buried, incinerated, or rendered or go into a secondary food consumption market. The incineration process itself can result in air pollution, and the burial process results in the same problems and/or benefits as manure.

The problems just outlined are not necessarily scale-specific. In theory, they arise from the concentration of animals in one place, regardless of how many ownership units are involved. It is possible that 1,000 animals belonging to 100 smallholders could create the same physical problems as the same number of animals belonging to one owner.

Environmental problems may occur in livestock production if producers follow at least one of the following practices: (1) directly dump manure, slurry, or processing water from livestock production into waterways; (2) stockpile undesirable byproducts in such a way that, as the nutrients go through nutrient cycles, components volatize into the air;[1] (3) fail to credit the nutrient content in this organic source and thus overapply it as a soil amendment in conjunction with chemical fertilizers;[2] (4) apply manure at the wrong time of the growing season; (5) apply manure in areas where the hydrogeomorphic profile is such that it is difficult to prevent run-off (areas with high water tables, sandy soils, or sloping terrain); (6) base land applications solely on the nitrogen requirements of the crop, which may result in overapplication of phosphorus (also found in manures); (7) use inadequate technology or otherwise fail to address hazards until problems arise due to storms and run-off; or (8) choose inappropriate methods of disposing of the carcasses of dead animals.

Environmental Externalities and the Rationale for Government Intervention

As discussed in Chapter 1, externalities occur when an economic agent reaps a benefit from an activity whose cost is born in part by others, but they are not compensated for this. Agents "capture" benefits from negative externalities if they receive a benefit (for instance, livestock sales) but someone else bears part of the cost of securing that benefit (in the form of odors, flies, polluted water, and so on) and is not compensated for that cost. Agents "internalize" at least some of these negative externalities to the extent that they themselves suffer from these ills, and also if they incur expenses to compensate those who bear the cost or to prevent the bad side effects. If farms of different sizes capture benefits from negative externalities that differ per unit of output across farms, those with larger capture per

unit of output gain a cost advantage relative to those who do not. If such differences vary systematically by scale, environmental externalities could provide an incentive for scale-up, or might in fact slow down scaling-up, depending on how they go.

There are four potential sources of negative environmental externalities in livestock production and processing that will cause problems over time in the absence of effective polices and institutions to deal with them. The first source of externalities has to do with manure. In the study countries, manure is traditionally spread directly on farmland as a nutrient source and soil enhancer, used as fish food, or used as a source of fuel. Currently, none of the studied countries has regulations requiring or otherwise regulating the application of manures, but Brazil, the Philippines, and Thailand do require swine operations (particularly large-scale operations) to have production permits. Extension agents in the survey areas of these three countries reported nutrient loading.

The second source of externalities concerns the disposal of dead animals. Currently, none of the study countries has regulations regarding dead animal disposal practices. A third source of externalities is embedded in the release of ammonia, which in zones of high animal density can be a source of air pollution, producing acid rain and/or odor problems. Both the Thai and Brazilian country studies report that farm neighbors have complained about odors; in Thailand, producers were forced to compensate neighbors for the smell (Poapongsakorn et al. 2003).

A fourth externality has to do with the discharge of wastewater from processing plants. Brazil, the Philippines, and Thailand have regulations concerning disposal

---

[1]For instance, some of the nitrogen in uncovered manure would volatize into the air as it went through the nitrogen cycle.

[2]In all the countries covered in this study, the animal manures are considered "organic," thus safe, and are applied to crops on top of chemical fertilizers, particularly in areas where the nutrients are not credited.

of this sludge, which in effect attempt to force companies to internalize this negative externality.

In principle, governments will improve the social good by intervening when there are externalities or imperfect information to better equate social costs and benefits. Intervention requires a clear definition of the property rights associated with the market failure being corrected. The economic rationale behind policy intervention would be to get private individuals to incorporate the full costs (including environmental costs) of production back into their decision-making process. Brazil, the Philippines, and Thailand have implemented environmental regulations, which will be discussed later in this chapter concerning livestock, but enforcement remains the problem.

Nevertheless, there are a variety of reasons why a government may have difficulty intervening or controlling livestock pollution. First, economic reasons may not be paramount in policymaking.[3] If this is the case, it could mean either that powerful large firms get away with more than small ones or, conversely, that it is more appealing to the government to go after one or two large farms while leaving problems created by tens of thousands of small ones untouched.

Second, government regulators also have to deal with information and attribution issues. There is always the non-point-source problem: regulators may not be able to determine who caused a specific problem. Furthermore, although a given livestock farm may be the source of the nutrients involved in pollution, regulators may have limited control over what happens once the manure leaves the farm, say, for sale to another farmer. All four of the country studies indicate that there is an active market for poultry manure. Yet none of the four countries has regulations concerning the spreading of manures on land.

Government failure to intervene may also have to do with the changing structure of the livestock industry and the fact that production systems are increasingly becoming integrated across farms. Specifically, as there has been a move toward specialized farms in both crop and animal production, there have also been changes in the nutrient balances of agricultural regions, soils, and individual farms. Specialization across crops on the one hand, and across livestock on the other, concentrates soil nutrients and increases feed and transportation costs. It also leads to high ratios of livestock per unit of land in certain areas, causing nutrients coming from animal waste to be applied at higher rates than required by crops. Stoyke and Waibel (1994) suggest that growth of this type of specialized farm in many countries was facilitated by the absence of regulation with regard to the number of animals per unit of farm area, rather liberal animal protection laws, and the availability of low-price imported feed stuffs.

The splitting of the custody of animals from ownership, as in the case of contract farming, complicates the issue of finding one party responsible for pollution. Brazilian swine integrators, who have contracts with a large number of small producers, are starting to recognize the potential legal liability. They have written demands for internalizing potential negative environmental externalities into their contracts with growers. These contracts include stipulations as to the adequate handling of animal waste. All the integrated farmers have to go through an environmental authorization process led by the integrator. The costs involved in this process are discounted from the payments to growers. Thus, growers bear the costs, but integrators make the decisions while potentially limiting their own liability for pollution created in the production of the animals they market. This opens scope for

---

[3]Because of the nonpoint nature of pollution created by poor animal waste management, the cost of monitoring can be excessive if the government implements specific control measures for each farm.

new externalities that benefit integrators at the expense of growers. In contract farming, integrators tend to be the ones that provide the inputs into production to the growers, and they are interested only in removing the live animal. Manure and dead animals tend to be left for the producer to dispose of (properly or not). The nutrients from these products can be viewed as either a positive (that is, as compost for the contractor's farm) or a negative externality, depending on how they are disposed of. In countries where the responsibility for these outputs is being questioned due to increased environmental regulations, integrators are beginning to write that responsibility into the contracts of their growers.

Increasingly, it is being recognized that there are economies of scale associated with control options. Schwart, Holt, and Outlaw (1995) have noted that there are economies of size not only in production technologies, but also in manure management, which reinforces the trend toward large production units. There are high start-up costs to build some types of manure storage systems and dead animal disposal systems. The marginal cost for extra pollution control is lower as producers increase in size. Thus, in theory, large producers have an advantage over smallholders in implementing animal waste control measures due to the number of animals they raise.

Moreover, the costs of improved manure management may fall disproportionately on broiler and dairy producers, who tend to concentrate in areas classified as environmentally vulnerable (Westenbarger and Letson 1995).[4] Producers in concentrated areas tend to have limited land on which to dispose of the waste, because often

the land has become saturated with nitrogen and phosphorus over the years, while in less concentrated areas, the land is not saturated. Producers are often located in concentrated areas close to the urban demand centers. Urban areas are already the source of many nonagricultural forms of pollution. Thus, production sites near urban areas have often become environmentally fragile over time. Further, integrators often prefer their contract growers to be concentrated to save on logistic costs, which further places a burden on some of these already environmentally fragile areas.

Policies and Institutional Mechanisms to Mitigate Environmental Pollution Problems Stemming from Livestock

Countries that appear to be having problems have implemented policies such as (1) licensing policies that restrict locations where certain amounts of pollution may be discharged, (2) clean water and air legislation, (3) regulations on disposal of poultry by-products, (4) limits on the number of animals per area, and (5) policies on the improvement of markets for poultry waste. Although many are just beginning to be implemented, it seems likely that most of them will retard overall growth in output as production becomes more expensive. If enforcement differs by scale of farm, it will also affect the size distribution of farms. If enforcement differs by locality, it will also shift production among locations.

To prevent water quality problems from rapidly growing livestock production, many countries have developed regulations con-

---

[4]Fallert, Weimar, and Crawford (1993) have shown that in the United States there are economies of size in waste management. According to estimates done by Outlaw et al. (1993), the cost to comply with waste management rules would be $400 per cow from a 50-head farm, compared to $288 per cow from a 175-head farm. Furthermore, the results from the study indicated that there could be a restructuring of the industry in response to proposed standards for confined feeding operations developed by the U.S. Environmental Protection Agency's Regional Office 6. The study also noted that although larger dairies would be able to assume the extra capital costs, smaller producers would require several years to recover their investments.

cerning the application of either nitrogen or phosphorus. The premise behind this trend is that the rate of animal waste applied should not provide more plant-available nitrogen than is required by the crop in question in order to avoid the contamination of groundwater by nitrates (NO_3). Studies have indicated that nitrate and phosphorus contamination in groundwater in areas with high animal densities is due to the unfavorable ratios of nitrogen to phosphorus in animal wastes relative to the uptake of these nutrients by crops. This imbalance results in the excessive phosphorus in the soil. An environmental problem will develop if there is a transport process that redistributes phosphorus that originated in animal waste to a surface water body sensitive to eutrophication.

Various rules and regulations have been developed in the different study countries to limit potential environmental problems from livestock. Brazil, the Philippines, and Thailand have the most comprehensive set of rules of the four countries (discussed in the following subsections). India has minimal environmental rules regarding livestock.

Thailand

In Thailand, the National Environment Quality Act (NEQA) was first enacted in 1969 as a comprehensive package for institutionalization of environmental policy and planning. The act was amended in 1992 as the Enhancement and Conservation of National Environmental Quality Act. Following passage of the NEQA, the Thai cabinet made a resolution in 1996 to adopt the "Policy and Plan for National Environmental Quality Preservation and Promotion (1997–2016)" (PEQP). Under PEQP, the line ministries and every province have to come up with their own respective action plans. PEQP has produced yet another action plan called the Environmental Quality Management Action Plan, which treated water pollution as its top priority. Specifically, water quality standards were set for two periods, 2001 and 2006, for the Chao Phraya River and the Tha Chin River. Regulations were announced in February 2001 with respect to the dumping of waste into watercourses, including effluent standards for pig farms. The new standards became effective on February 24, 2002. Thus far, the standards have been monitored and enforced only on large- and medium-sized farms.

Brazil

A series of strict laws at both the federal and state levels have been passed in Brazil since 1965 to aid in the general protection of the environment, and they all have ramifications for the livestock sector, especially large-scale enterprises (Camargo Barros et al. 2003). The original and most significant is Law 4771 on forest management, passed in 1965 and amended in 1989 and 1996. It protects areas around rivers, lakes, springs, and dams as well as any area surrounding watercourses.

A series of other measures on water use and protection have also been passed. As of this writing, environmental legislation is being discussed in the Brazilian congress. The River Basins Committee, whose members are municipalities, nongovernmental organizations, industries, farmers, and public offices, is an instrument created by the federal government (now being adopted by state governments) that gives small producers access to environmental discussions. Its function is to discuss and make decisions about subjects related to the management of water resources in one or more river basins. Financial institutions have also become involved in environmental issues through the implementation of credit and financing policies for the agricultural and urban zones. The Environment Conservation and Control Financing Program—"FNE Green," implemented through state-run development banks for the benefit of rural industry and agroindustries (persons and companies), cooperatives, and associations—has the objective of promoting the development of productive environmental activities and the financing of conservation measures.

The Philippines

Environmental regulations issued periodically from 1976 to 2001 to control pollution are quite stringent in the Philippines. A 1978 law is in fact patterned after the U.S. Environmental Impact Statement System (Costales et al. 2003). Environmental laws on pollution in the Philippines originally targeted the regulation of pollution from heavy and light manufacturing and assembly industries. These standards were initially applied only to commercial livestock farms with more than 1,000 animals in their inventory (Catelo 2002). However, because smallholders constitute close to 80 percent of hog inventories and because backyard production in the major producing towns, provinces, and regions is also densely configured, the issue of pollution from these operations has since been raised. The rationale is that these farms must create as much potential pollution per animal as commercial farms, and they are now concentrated in commercial densities in periurban areas. If this is so, the thinking goes, they too must be regulated. The question of whether large-scale farms pollute more than small-scale farms is a pressing policy issue for the Philippines that has yet to be empirically resolved.

Enforcement Problems and Differential Impacts of Enforcement on the Scale of Production

There was a lack of information on enforcement issues from India at the time of the study; hence, only enforcement problems from Brazil, Thailand, and the Philippines are discussed here.

Thailand

In Thailand, most attention has been paid to large swine farms, in terms of both regulation and enforcement. All pig farms with more than 500 sows have invested in water treatment, while nearly 13 percent of small-scale farms have no sewage treatment system at all (Poapongsakorn et al. 2003). In terms of treatment techniques employed by farms of different sizes, all of the large-scale farms (with 5,000 pigs or more) use a pond system or a solid-liquid system to treat their sewage. Medium-scale and small-scale farms tend to use a single reception pond; this is not very effective in treating the sewage, because the water that is spilled over to the drainage channel is still much polluted. Mainly small and medium farms use biogas digesters. It is probable that many large farms do not find this technique cost-effective, because many are new farms settled in land-ample areas where they can make do with pond systems.

Although a number of rules and regulations on wastewater management have been developed in Thailand and some incentives are offered to livestock farms for investing in waste treatment technology, there has been a lack of enforcement. The Department of Livestock and Development (DLD) is one of the inspection offices and is expected to be a key institution to enforce the environmental protection law. Ironically, DLD's main job is extension, and it serves as a support unit for livestock farmers. Its acting as an inspection office might cause conflicts between the department and farmers; this would lead to difficulties in working with farm owners. Consequently, DLD may not be able to act efficiently as an inspection agency. Thus, more effective enforcement and implementation of existing laws are needed to achieve responsible environmental management.

Brazil

While applicable laws exist in Brazil, there are problems with the enforcement of these laws due to the limited number of personnel relative to the number of farms and the lack of agreement from farmers with government policies in this area. In a 2001 survey of 3,505 agricultural producers, 76 percent considered environmental issues a problem that must be managed with the application of technical regulations, scientific knowledge, and skilled personnel. They recognized the importance of promoting environmental conservation, observing, however, that the procedures must be based on technical knowledge. Yet only 5 percent of the producers approved the government's current environmental plan.

As the livestock sector has become more industrialized, livestock farms have tended not only to be larger in size, but also to be run more as big businesses. As such, they tend to be under closer scrutiny by national, state, and local authorities than are farms owned by smallholders. Moreover, there may be a policy bias in terms of enforcing regulations against large firms first, in contrast with the standard "political economy" view that the rich get away with more than the poor.

The Philippines
The Philippines has begun to recognize the growing problem of pollution from smallholder livestock. Regulatory agencies that have cracked down on large farms in the past have begun contemplating issuing regulations on pig waste disposal by small farms. One such agency in Southern Luzon is the Laguna Lake Development Authority (LLDA) in the high-density Metro Manila livestock zone. In addition to spearheading the management of the water resources of the Laguna de Bay, the largest freshwater lake in the country, the LLDA also functions as a special environmental authority. As such, it has regulatory and juridical functions in the enforcement of environmental laws and the issuance of penalties on firms violating environmental standards for the lake. Very recently the LLDA issued Resolution 169 approving policy guidelines governing the operation of backyard/small-scale hog farms in the Laguna Lake region.

Capture of Environmental Externalities
Though the information already given is of interest, it tells of little difference across

size of farms in their ability to internalize the environmental externality associated with manure and dead animal disposal. Thus, it does little to aid us in testing Hypothesis 3, *Small farmers expend a greater amount of effort/investment in abatement of negative environmental externalities per unit of output than do large farmers,* and in understanding why this may be the case.

To look at this issue in detail, two measures are used: (1) the ability of different-sized farms to assimilate all the nutrients produced on-farm in terms of the estimated mass balance of nutrients produced from the manure based on the number of animals and the acreage across different-sized farms and (2) expenditure differences across different-sized farms to mitigated negative environmental effects from manure and dead animal disposal. Calculations are made of the relative contributions of large and small farms to excess nutrient balances in the study countries. Large and small farms are also compared with respect to the estimated money value of their efforts to mitigate negative environmental externalities using the approach laid out in Chapter 3. These two methods provide both a direct and an indirect approach to the question of which pollute more per unit of livestock output in the study countries: large or small farms.

Evidence on the Impact of Increasingly More Concentrated Livestock Production on Excess Mass Balances of Nutrients
As illustrated in Chapter 2, most livestock production in the study countries is concentrated in and around major cities or areas otherwise favored by infrastructure. This concentration can lead to massive surpluses of nutrients (nitrogen, phosphate, and potassium) in these areas, causing pollution and eutrophication of surface water and groundwater (see http://www.fao.org/WAIRDOCS/LEAD/X6170E/x6170e16.htm for an illustration of nitrogen and phosphorus loads in Asia in 2002). The emission of greenhouse (methane, nitrous oxide) and other

gases (ammonia) is another important form of pollution.

Disposal practices for manure and dead animals directly affect nutrient balances and also vary across scales of operation. Traditionally, byproducts of livestock production in the four countries studied, as found through the surveys, have been disposed of by (1) applying them to land to improve the physical properties of the soil for plants, (2) using them as animal feed (for other animals, including fish and oysters), (3) using them to create energy (directly burning them for heat or the creation of methane gas), (4) using them as a cement lining for floors, (6) dumped them into rivers, (7) burying them, (8) incinerating them, and (9) using them for brick-making. Though all approaches are used, the most common are land application, energy production, and dumping.

Swine production is probably the least environmentally friendly form of livestock production of the types of production looked at in this study. As a result, countries such as Brazil, the Philippines, and Thailand have set regulations to control sewage from these operations. Brazil and Thailand are also providing funds to aid farmers in protecting the environment. For poultry, industrialization is less obviously an environmental problem. This is partly because, in all the study countries, chicken manure can be sold. Therefore, the increased volumes of manure produced by the large farms have not created the same absorption problem as in swine production.

In Thailand, manure coming from poultry farms goes into a holding pond, cesspool, or biodigester or is dried. It may then either be sold to a middleman or used on a farm as a fertilizer or as fish food. Dead animals are used either on-farm or sold to a secondary market for food or human consumption.

In India, broiler manure tends to be removed from animal housing by either family or hired labor. It is then piled in either an open or a closed shed or used immediately.

If it is used on-farm, it is used primarily for organic fertilizer. If it is not used on-farm, most likely it will be sold for organic fertilizer, brick-making, fuel, or mushroom substrate, or it may be dumped. Similar practices exist in other study countries, but usually in these countries broiler manure is valued as fertilizer.

In Brazil, layer manure is either used on-farm or sold to manure trades. Dead animals are put into a cess pool, composted, or buried. In the Philippines, manure cleaned out of the houses may go into an open pit or be laid on the ground. After that, it will be either used on-farm or sold to the market. Dead piglets tend to be buried. Other countries may have lagoons for their manure, use biodigesters, or feed it to fish.

Under these disposal practices, the value of manure can vary from a net financial cost to the farm when the material is hauled away and destroyed to a valuable product of the farm when it is sold for use on fruits and vegetables. Various entrepreneurs have developed ways to add value and improve manure use, such as marketing it as an enhanced fertilizer, an animal feed, or fuel. The success of the marketing options depends on the types of crops using the manure, the distance from the farm that needs it, and the willingness of the buyer to use the manure as a product.

The potential buyers of manure are other farmers (for fruit, vegetable, and field crops, particularly for organic farming), commercial nurseries, home and garden centers, greenhouses, and even homeowners. In most cases, the market for manure is very local because of the cost of transportation. Undesirable characteristics associated with raw manure are being too wet, lacking uniformity, containing too many feathers, possibly containing toxic chemicals, often being lumpy, containing weed seeds, yielding an unpredictable response, often having the wrong balance of nutrients, not always being available when needed, providing a good breeding ground for flies, possibly burning the crops, and being smelly.

Because ruminants are able to use the nonprotein nitrogen in poultry and swine wastes, the latter can also be used as a supplement feed for certain animals. Furthermore, both Thailand and India reported the use of some manures as fish feed. The acceptability of the use of manure and dead animals as feed additives is dwindling worldwide because of concern about bovine spongiform encephalopathy (BSE), or "mad cow" disease. Although Brazil previously used meat and bone meal as feed additives, it has stopped doing this since both the E.U. and the United States have banned imports from countries permitting ruminant-to-ruminant feed practices. Even use in feeds for nonruminants has ceased because of the danger that Brazil will be decertified as an exporter.

Another potential market for manure is the production of methane through anaerobic digestion. It can be derived directly through a biological process such as combustion or indirectly through generation of biogas, which is a mixture of methane and carbon dioxide. Most methane produced has been burned for heat or used as a fuel for internal combustion engines. The equipment used for producing methane is not simple and is fairly expensive, posing a problem for the economic feasibility of biogas production. The major factors in the high costs of these methane-producing structures are the high cost of properly designed structures, mixing equipment, and gas-control devices; the precautions necessary to avoid explosions; and the need to store methane and the digested liquid manure.

The generation of biogas has multiple benefits besides the production of a combustible fuel. For instance, byproducts from sludge of a digester that uses manure can be used as a feed supplement, while the liquid effluent can support fish growth and make aquaculture possible. The economic feasibility of methane production is increased

when the options for use of the residues are also taken into account. The problem with this is that after methane has been generated, one still has to get rid of the excess nitrogen and phosphorus.

Because of the expense of biogas generation, Thailand has started programs that provide partial funding for construction of waste treatment systems such as biogas digesters and multilevel ponds. The amounts provided by the Thai government for installing wastewater treatment systems range from 20,000 to more than 150,000 baht per farm, approximately 40–60 percent of the total installation cost.[5] The Thai government intends to subsidize up to 38 percent of the construction cost, provide low-cost loans, and facilitate loan approval for farms.

Farms with insufficient land for spreading manure or facilities for otherwise reducing it can still improve mass nutrient balances through sales of manure to others. Presumably there is a market for manure only because of its value as a fertilizer or fuel, so manure sales and transport off the farm are generally good signs for management of mass balances.

Current Pathways for Use of Nutrients from Manure and Dead Animals

Figures 6.1–6.3 summarize the various pathways for nutrient management for poultry, swine, and dairy operations currently used by the sample farmers in the four countries. Poultry operations tend to use the manure for either crops or fish feed, and dead animals tend to be buried, incinerated, consumed, or used as fish feed. After it is cleaned out of bird enclosures, poultry manure in most of the study countries tends to be dried and sold to be used on crops or as fish feed in the Asian countries. If it is not sold, it remains on the livestock farm and is used for the same purposes. The use of swine manure differs according to whether

---

[5]Approximately 42 baht equaled US$1 at the time of the survey in 2002.

Figure 6.1   Pathways for disposal of poultry manure and dead birds

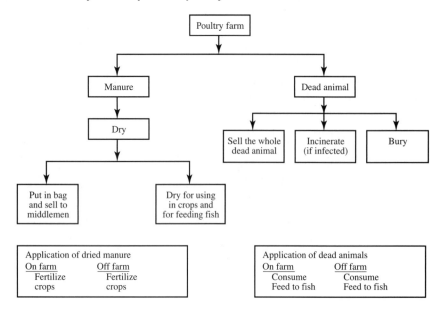

Source: Poapongsakorn et al. (2003).

Figure 6.2   Pathways for disposal of swine manure and dead piglets

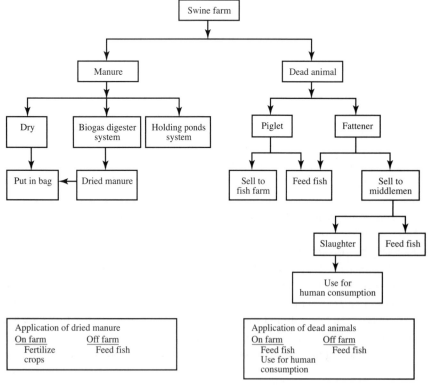

Source: Poapongsakorn et al. (2003).

Figure 6.3   Pathways for disposal of dairy manure and dead animals

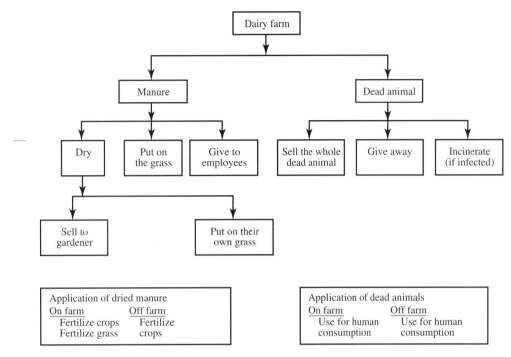

Source: Poapongsakorn et al. (2003).

it is in wet or dry form. Wet manure may be used to create household and farm energy, such as biogas for cooking or power for farm use, as in Thailand. Dry manure is either applied to fields as a fertilizer or used as a feed in fish farming in the Asian countries studied. In Brazil, though the use of manure as fish food has been reported in the past, it currently is not used due to the impression that it results in a lower feed conversion than palletized feed. The manure and wastewater created on dairy farms will typically be left on the grass where it falls, given to employees, or dried. If dried, it may be sold to a neighbor, put on the farmers' own land, or burned for fuel.

Dead animals of all types of livestock in India, the Philippines, and Thailand may be sold to a secondary market for either human or animal consumption, given away, buried, or incinerated. The use of dead animals is not reported in Brazil, which has banned

their use for fear of BSE and potentially lost markets in countries that prohibit the use of rendered products.

Manure Disposal for Survey Households

Tables 6.1–6.4 summarize how households in the study countries chose to dispose of the manure generated from their livestock operations. For broiler manure, the majority of the households chose to get rid of it through the market. A slightly lower percentage of small-scale producers across the households surveyed in all the countries sold the broiler and layer manure, mainly because it appears that they were using it on their own farms. This is not unexpected, because many small-scale farmers are mixed-product farmers and can readily use the manure as an organic fertilizer.

In India, 79 percent of the small-scale broiler producers and 95 percent of the

Table 6.1 Distribution of farmers by manure disposal methods, broiler production, 2003 (percent)

| | Small (10,000 or fewer birds) | | | | | | Large (more than 10,000 birds) | | | | | |
| | Brazil | India[a] | Philippines | | Thailand[b] | | Brazil | India[a] | Philippines | | Thailand[b] | |
Disposal method	Contract		Independent	Contract	1–5,000	5,001–10,000	Contract		Independent	Contract	10,001–20,000	More than 20,000
On farm		14		10	46	20		5	4	3	15	11
Fish farm												
Crops	44						33					
Off farm												
Sold	32	79	36	65	46	63	49	95	61	81	78	67
Gifted		7										
Used both on farm and off	16		7	3	8	18	16		13	16	7	22
Noneconomic use												
Dumped in waterways			10									
Laid on ground			19	3					4			
Other	8		28	19			2		18			
Total	100	100	100	100	100	101	100	100	100	100	100	100

Sources: Compiled from Camargo Barros et al. (2003); Costales et al. (2003); Mehta et al. (2003); Poapongsakorn et al. (2003); Sharma et al. (2003).

[a]For India, the definition of the small (large) category is fewer than 10,000 birds (10,000 birds or more).

[b]Thai farms have multiple responses, so percentages exceed 100 percent.

Table 6.2   Distribution of farmers by manure disposal methods, egg production, 2003 (percent)

| | Small (10,000 or fewer birds) | | | Large (more than 10,000 birds) | | | |
| | | | | | | Thailand[b] | |
Disposal method	Brazil	India[a]	Thailand[b]	Brazil	India[a]	10,001–20,000	More than 20,000
On farm		10			10	9	91
Fish farm			41				
Crops			20				
Other			29				
Off farm							
Sold	67	90	37	72	90	63	4
Gifted							
Disposed							
Dumped							
Used both on farm and off	33		7	28		28	13
Total	100	100	134	100	100	100	108

Sources: Compiled from Camargo Barros et al. (2003); Costales et al. (2003); Mehta et al. (2003); Poapongsakorn et al. (2003); Sharma et al. (2003).
[a]For India, the definition of the small (large) category is fewer than 10,000 birds (10,000 birds or more).
[b]Thai farms have multiple responses, so percentages exceed 100 percent.

large-scale producers sold manure in the marketplace. Most of the manure was piled in either open sheds or closed sheds. When used on-farm, poultry manure was used primarily as an organic fertilizer. When used off-farm, poultry manure was used as a fertilizer or for making building bricks, fuel bricks, or mushroom substrate.

In the Philippines, the most common scenario among the farms surveyed was disposal of manure by selling it to chicken manure traders, spreading it on their own farms, or some combination of both. Other observed means of disposing of waste were containing the manure in closed pits, throwing it into a river or canal, or just leaving this byproduct in piles on the ground to decompose. In the Philippines, a larger percentage of the contract farming smallholders sold the manure than did the independents (Table 6.5). This may reflect that they

tend to have less land per animal than do independents, and they are also less likely to be cropping. If they did, they might not be selling the manure but might be using it as a source of nutrients for growing feed.

In Thailand, a high percentage of larger size egg producers sold their manure than did smaller operators. This probably reflects the fact that the larger farms have grown quite large relative to the limited land they have for disposal of the manure. For instance, 91 percent of the large-scale layer operations in Thailand sold manure, compared to only around 30 percent of the small-scale producers.

In Brazil, a lower share of manure was sold off-farm than in the Asian cases, and the price of manure differed by state. Table 6.6 shows that in the south, an area with highly concentrated poultry and swine operations, manure is cheaper due to its relative

Table 6.3 Distribution of farmers by manure disposal methods, swine production, 2003 (percent)

Disposal method	Small (1–100 head)				Medium (101–1,000 head)							Large (more than 1,000 head)				
	Brazil	Philippines[a]		Thailand[b]	Brazil			Philippines[a]		Thailand[b]		Brazil		Philippines[a]		Thailand[b]
		Ind	Contract		Co-op	Ind	Contract	Ind	Contract	101–500	501–1,000	Co-op	Ind	Contract	Commercial	
On farm																
Crops		21	4		100	100	87	23	23			74	73	73	23	
Biogas		7	4	5					5	13	14				7	12
Off farm																
Sold		1		50			13	1	5	72	69	3	4	3	3	93
Gifted				15						3	10	23	23	24		5
Used both on																
farm and off		1						4	2	1					3	
Biogas								9								
Noneconomic use																
Fish farming				5						1	7					7
Thrown in canal/river		3	9	25						6	1					1
Laid on ground		15	13													
Put in open pit		20	30													
Put in lagoon		20	30					62	65						63	
Put in septic tank		13	9					1							1	
Not reported	100									4						
Total	100	101	99	100	100	100	100	100	100	100	101	100	100	100	100	118

Sources: Compiled from Camargo Barros et al. (2003); Costales et al. (2003); Mehta et al. (2003); Poapongsakorn et al. (2003); Sharma et al. (2003).

Note: co-op, cooperative; ind, independent.

[a] Total percentages for Philippine farms may not add up to 100 percent due to rounding errors.

[b] Thai farms have multiple responses, so percentages exceed 100 percent; details of Thai wastewater are presented in Poapongsakorn et al. (2003).

Table 6.4 Distribution of farmers by manure disposal methods, milk production, 2003 (percent)

| | Small | | | Medium | | | Large | | | Commercial/periurban |
| | 50 or fewer | 1–4 | 1–20 | 51–70 | 5–10 | 21–30 | More than 70 | More than 10 | More than 50 | |
Disposal method	Brazil	India	Thailand	Brazil	India	Thailand	Brazil	India	Thailand	India
On farm										
Fertilize crops			13			7			16	
Fertilize grasslands			13			31			37	
Used as fertilizer for both	100		3	100		2	100		5	
Off farm										
Sold			29			24			5	
Gifted			3			2			5	
Sold and gifted						2				
Used both on farm and off			31			27			32	
Used as fertilizer for both		30			55			68		84
Used as fuel		70			45			32		16
Noneconomic use										
No response			8			5				
Total	100	100	100	100	100	100	100	100	100	100

Sources: Compiled from Camargo Barros et al. (2003); Poapongsakorn et al. (2003); Sharma et al. (2003).
Note: No response covers farmers who were part of the sample but did not respond to the survey.

Table 6.5 Manure sales from broiler production in the Philippines, 2003

| Variable | Smallholder (10,000 or fewer birds) | | Commercial (more than 10,000 birds) | |
	Independent (N = 13)	Contract (N = 21)	Independent (N = 17)	Contract (N = 30)
Manure sales (percent)	41.0	67.7	73.9	96.7
Volume of manure (bags)	83	289	38.1	4.24
Value of manure				
Pesos/batch	1,293	4,362	3,538	4,657
Pesos/100 kg output	86	23	9	14

Source: Costales et al. (2003).

Table 6.6 Average price of manure by Brazilian state, 2003 (reals/ton)

State	Price of manure
Paraná	14.37
Santa Catarina	12.04
Rio Grande de Sul	6.42
Mato Grosso do Sul	29.29
Mato Grosso	34.38
Goiás	45.29
Minas Gerais	61.08
Mean	28.98

Source: Camargo Barros et al. (2003).

abundance. In the center-west and Minas Gerais, the value of manure is almost 10 times the price in Rio Grande do Sul, an area with heavy concentrations of poultry and swine. For the most part, over 50 percent of broiler manure was sold off-farm. The exception was among smallholders, whose use on-farm and off-farm was about evenly divided. Most of the manure was sold directly to other farmers, but a small percentage was sold to wholesale traders, with the share going to wholesale trader increasing from 5 percent for the small farms to 19 percent for the large.

Dead Animal Disposal by Survey Households

Tables 6.7–6.9 summarize the percentage of households across countries and sizes of operations using specific disposal methods for dead animals. The specific method differed by country and size of operation. Ninety percent of the small independent swine producers in the Philippines buried dead piglets. The majority of large-scale producers, on the other hand, disposed of dead animals through incineration or some other method. Close to 90 percent of the broiler farms surveyed disposed of dead animals within the farm premises. The animals were buried, burned, thrown into pits, or fed to other animals. In Thailand, dead swine often

were given away as gifts, sold to the market, or sold back to the contractor. These practices were fairly uniform across sizes of operations. The Philippines is the only country to report that a small percentage of the households (1 percent) dump the dead piglets into a river. This may be due to a requirement of the integrator to ensure that the pig died of natural causes.

Similar results were found for broilers and layers. As with swine, most of the households surveyed disposed of dead animals on-farm. The most common means of disposal of dead broilers (Table 6.7) for smallholders in India was burial (70 percent), followed by sale to a secondary market (27 percent). A much smaller percentage of the large households in India reported this practice, but there were too many "no" responses in this category to verify that assumption. In the Philippines, around 30–35 percent of all producers dispose of birds through incineration, and 10 percent of these are small independent producers. In Thailand, incineration and burial were reported together, and these practices appear to be the most popular among the households surveyed.

For layers (Table 6.8), the preferred place to dispose of dead animals is on-farm for small-scale producers in both India and Thailand. More than half of India's small-scale producers chose burial over incineration. This may be due to the higher capital cost associated with incineration opposed to burial. A relatively large percentage reported disposal in the "other" category. When the enumerators were doing the survey, they found refrigerators at the back of many farm houses used to store dead animals, and it appears that these animals are often sold at a reduced price on a secondary market. This practice was also reported in Thailand. Fifty-eight percent of the large egg producers surveyed in Thailand dispatch their dead animals to retailers.

In India, there was similar mortality in the grow-out phase for small-scale producers and large-scale producers (4 percent

Table 6.7 Distribution of farmers by dead animal disposal methods, broiler production, 2003 (percent)

Disposal method	Small (10,000 or fewer birds)						Large (more than 10,000 birds)					
	India[a]	Philippines Independent	Philippines Contract	Thailand 1–5,000	Thailand 5,001–10,000	Brazil Contract	India[a]	Philippines Independent	Philippines Contract	Thailand 10,000–20,000	Thailand More than 20,000	Brazil Contract
On farm												
Buried	70	42	33				9	42	30			10
Incinerated	3	33	40	39	53		9	33	43	44	50	
Placed in open pit			10						7			
Placed in closed (cess) pit		13	13			3		13	10			24
Frozen for secondary market	27											
Fed to animals		13	3					13	10			
Fed to fish				28	6					7		
Composted						76						61
Other				5		21				7		5
Off farm												
Sold to fish farm				2	4					2	3	
Sent back to contractors				1	1					2	19	
Sold to retailers				1	1					3	6	
Other				3	9					4		
Mixture of on-farm and off-farm disposal				18	20					30	22	
Other				4	6		82					
Total	100	101	99	101	100	100	100	101	100	99	100	100

Sources: Compiled from Camargo Barros et al. (2003); Costales et al. (2003); Mehta et al. (2003); Poapongsakorn et al. (2003); Sharma et al. (2003).

Note: Total percentages may not add up to 100 percent due to rounding errors.

[a]For India, the definition of the small (large) category is fewer than 10,000 birds (10,000 birds or more).

Table 6.8   Distribution of farmers by dead animal disposal methods, egg production, 2003 (percent)

| | Small (10,000 or fewer birds) | | | Large (more than 10,000 birds) | | Thailand | |
| | | | | | | 10,001– | More than |
Disposal method	Brazil	India[a]	Thailand	Brazil	India[a]	50,000	50,000
On farm							
Buried		68	29	30	67	14	4
Incinerated	67	5	46	25	15	52	38
Placed in closed (cess) pit				44			
Composted	33			1			
Other		27			18		
Off farm							
Sold to retailers			25			33	58
Total	100	100	100	100	100	99	100

Sources: Compiled from Camargo Barros et al. (2003); Costales et al. (2003); Mehta et al. (2003); Poapongsakorn et al. (2003); Sharma et al. (2003).
Note: Total percentages may not add up to 100 percent due to a rounding error.
[a]For India, the definition of the small (large) category is fewer than 10,000 birds (10,000 birds or more).

mortality for both). Most dead animals were buried or sold to a secondary market as a source of food. Close to 70 percent of both small- and large-scale broiler producers buried their dead animals. Selling frozen dead birds to a secondary market was also practiced by small-scale producers (26 percent) and large-scale producers (19 percent). Incineration was practiced more by large-scale producers (9 percent) than by small-scale producers (3 percent). This may be due to the capital cost involved. Although feeding dead animals to fish is reported to be practiced in India, none of the households surveyed disposed of their dead birds in this manner.

Mass Balance Calculations

Figures 6.4–6.7 illustrate the percentage of farmers, by size category, within a given range of nitrogen nutrient balance. With reference to the discussion in Chapter 4, a range of –20 to –30, for example, indicates

that farmers are creating an annual excess of 20 to 30 metric tons of nitrogen (for example) per hectare. The mass balance calculations performed here are used as rough estimates of actual nutrient balances. They indicate households with potential problem areas, as well as where further research and technology transfer may be more productive for households of a certain size. They also indicate why some households have active involvement with manure markets.

The mass balance for swine production showed a similar range of excess nitrogen and phosphorus across countries. Figure 6.4 shows nitrogen mass balances (that is, land deficits, indicating nutrient loading) for swine production in Thailand (for mass balance results in Brazil and the Philippines, see Camargo Barros et al. 2003 and Costales et al. 2003). For each of these countries in which the swine-producing population was sampled, larger producers for the most part showed larger deficits, indicating a greater

Table 6.9 Distribution of farmers by dead animal disposal methods, swine production, 2003 (percent)

| | Small (100 head or fewer) | | | Medium (101–1,000 head) | | | | | | | Large (more than 1,000 head) | | | | |
| | Philippines | | Thailand | Brazil | | | Philippines | | Thailand | | Brazil | | | Philippines | Thailand |
Disposal method	Ind	Contract	Thailand	Co-op	Ind	Int	Ind	Contract	101–500	501–1,000	Co-op	Ind	Int	Comm	Thailand
On farm															
Buried	92	100	40	17	80	43	80	71	23	17	19	18	8	77	14
Incinerated	1						6	7						6	
Placed in open pit							9	23						2	
Placed in closed (cess) pit	4			17	20	7					43	55	15	13	
Consumed or fed to dogs	2		5				1		3					1	
Fed to fish			10				1		6	17					2
Composted				67		50					28	24	66		12
Off farm															
Dumped in river	1														
Sold as feed							2								
Gifted			10						9	5				1	2
Sent back to contractors									10	12					2
Other			5						3		9	2	11		
Not reported			30						46	49					67
Total	100	100	100	101	100	100	99	101	100	100	99	99	100	100	99

Sources: Compiled from Camargo Barros et al. (2003); Costales et al. (2003); Mehta et al. (2003); Poapongsakorn et al. (2003); Sharma et al. (2003).
Notes: comm, commercial; co-op, cooperative; ind, independent; int, integrated. Total percentages may not add up to 100 percent due to rounding errors.

Figure 6.4   Nitrogen mass balances (in metric tons) for swine farms surveyed in Thailand, 2002

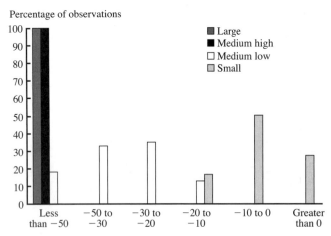

Source: Poapongsakorn et al. (2003).

Figure 6.5   Nitrogen mass balances (in metric tons) for broiler farms surveyed in Thailand, 2002

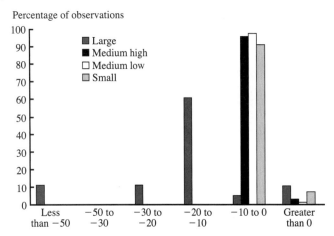

Source: Poapongsakorn et al. (2003).

Figure 6.6   Nitrogen mass balances (in metric tons) for broiler farms surveyed in India, 2002

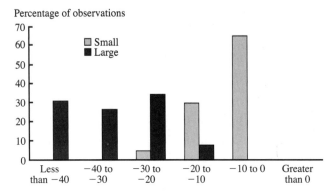

Source: Mehta et al. (2003).

Figure 6.7   Nitrogen mass balances (in metric tons) for dairy farms surveyed in India, 2002

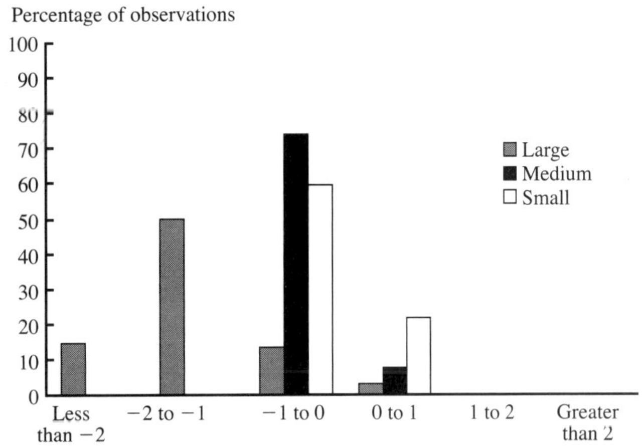

Source: Sharma et al. (2003), Annex IV.

degree of nutrient loading and thus greater need to find adequate disposal methods for manure. Each country has some households with sufficient land to assimilate the nitrogen produced. Smaller producers more often than larger producers had positive balances, indicating that further application of plant nutrients would be desirable. Brazil, with its larger amount of land availability, has households with positive balances in every size category. However, most of its large-scale producers did not have enough land to dispose of the manure based on the analysis here. Conversely, Thailand's large and medium-high producers all have large negative nutrient balances, suggesting a serious pollution problem.

Land nutrient balances for broiler producers show wide variation between countries (Figure 6.5; for mass balance results in Brazil, India, and the Philippines, see Camargo Barros et al. 2003; Costales et al. 2003; and Mehta et al. 2003). Large producers are likely to have large nutrient absorption deficits, while smaller producers are more likely to have small deficits. Small producers in Thailand (Figure 6.5) have the largest number of households with positive mass balances. Around 5 percent of Brazil's small and medium producers could absorb

manure in their own land. Both India (Figure 6.6) and the Philippines (see Costales et al. 2003) show no households with a positive balance. This underlines the high proportion of households producing on relatively small plots of land. All of the sample Philippine broiler producers had an excess of nitrogen nutrients from manure, so they had to worry about disposal of manure off site.

In sum, for smaller producers with lower excess nutrient balances, informal mechanisms to dispose of manure may be sufficient as long as they meet environmental requirements. In all the countries surveyed, there appears to be an active market for poultry manure to aid them in this. Large producers, on the other hand, typically require systems that guarantee that excess manure is disposed of in a controlled manner.

Relative to other livestock production, the mass balances for dairy production are comparatively in balance. Dairy production, by its nature, requires land. Thus, producers may have sufficient land to properly dispose of manure or may have access to nearby land. It is possible that small-scale producers in places like India (Figure 6.7) or Thailand (see Poapongsakorn et al. 2003) may not warrant sophisticated manure disposal mechanisms if they have adequate land.

Similar to the case of broiler production, the mass balances for egg production show wide variation between countries (Figure 6.7; for mass balance results in Brazil and India, see Camargo Barros et al. 2003 and Mehta et al. 2003). Large producers are likely to have large nutrient absorption deficits, while smaller producers are more likely to have small deficits. No households in Thailand (Figure 6.8) or India (see Mehta et al. 2003) have a positive balance. However, the excess of small producers in both countries is relatively small. Smaller egg-producing households with lower excess nutrient balances have informal mechanisms, such as an active market for poultry manure, to dispose of waste, and these may be sufficient, as in the case of small broiler farms. Large producers of eggs, as in the case of broilers, may require specific systems that guarantee that excess manure is disposed of in a controlled manner.

Average of Farm-Specific Measures of Environmental Mitigation Efforts by Farm Size

One of the key questions addressed in this study is the relationship between the size and internalization of environmental externali-ties per unit of output. Following the methodology set out in Chapter 4, each country study estimated a farm-specific index of the money value per unit of output of specific efforts to mitigate negative environmental externalities from manure and dead animal disposal. As outlined in Chapter 4, an aggregate figure per farm per commodity was first calculated, and this was divided by the total farm output of that commodity to get each farm-specific figure, and these figures were averaged across farms in each class. Only positive additions to mitigation were counted (failure to act assumes, de facto, that an externality is being created). Additions came from actual expenditures on compensating neighbors or collecting manure, building facilities to store it (suitably amortized), and disposal of manure at the sale price if actually sold or at the prevailing market price if spread on one's own fields. If no manure market existed, it is presumed that the manure supply is in excess and that applications to fields are not mitigating.

Table 6.10 shows environmental expenditures per kilogram of output for poultry. For broilers, smaller producers "pay" (internalize) more per kilogram of output in each country. The difference in internalization by

Figure 6.8   Nitrogen mass balances (in metric tons) for layer farms surveyed in Thailand, 2002

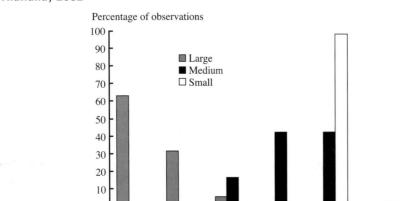

Source: Poapongsakorn et al. (2003).

Table 6.10   Environmental expenditures per kilogram output of liveweight broiler or eggs from poultry production, 2003

Poultry product	Farm size/type				
	Small (10,000 or fewer birds)		Large (more than 10,000 birds)		
	Independent	Contract	Independent	Contract	Total
Philippines					
Broiler (pesos/kg)	1.236	0.313	0.128	0.223	0.500
Number of samples	31	31	23	31	116
	Small (fewer than 10,000 birds)		Large (10,000 or more birds)		
India					
Broiler (rupees/kg)	2.548		0.397		1.880
Number of samples	92		43		135
Layer (rupees/egg)	0.0004		0.0002		0.0003
Number of samples	63		98		161
Brazil					
Broiler (reals/kg)		0.014		0.006	
Number of samples		34		201	
Layer (reals/kg of eggs)	0.100		0.040		
Number of samples	6		83		
	1–5,000 birds	5,001–10,000 birds	10,001–20,000 birds	More than 20,000 birds	
Thailand					
Broiler (baht/kg)	0.008	0.005	0.001	0.02	0.007
Number of samples	74	51	27	18	170
	10,000 or fewer birds		10,001–50,000 birds	More than 50,000 birds	
Layer (baht/egg)	0.06		0.05	0.03	0.05
Number of samples	41		32	23	96
Total variable cost (TVC)	1.19		1.14	1.14	1.17
Percentage share to TVC	5.0		4.4	2.6	4.3

Sources: Compiled from Camargo Barros et al. (2003); Costales et al. (2003); Mehta et al. (2003); Poapongsakorn et al. (2003); Sharma et al. (2003).

farm size is striking. In each country except Brazil, smaller producers internalize, on average, over five times more per kilogram in environmentally related cost. In Brazil, the difference is a factor of two. For layers, the relative difference is less. For Brazil and Thailand, internalized environmental costs for egg-producing households are higher on average than for broiler-producing house-

holds. In India, the opposite holds. In the Philippines, small contract producers internalize less per kilogram than independents, but larger contact producers pay more than large independents. The Brazilian layer costs seem much lower relative to the broiler costs. The reason needs to be further explored.

The internalized environmental costs per kilogram of output for swine are shown

Table 6.11 Environmental expenditures per kilogram output from swine production, 2003

Variable	Farm size/type					
	Small (100 or fewer head)	Medium (101–1,000 head)			Large (more than 1,000 head)	Total
Brazil						
Expenditure (reals/kg)	0.034				0.028	0.03
Number of samples	31				161	192
		101–500 head	501–1,000 head			
Thailand						
Expenditure (baht/kg)	1.20	1.02	0.76		0.46	0.840
Number of samples	20	70	41		43	174
Total variable cost (TVC) (baht/kg)	39.53	27.54	24.99		21.07	26.72
Percentage share to TVC	3.0	3.7	3.0		2.2	3.1
	Independent	Contract	Independent	Contract	Independent	
Philippines						
Expenditure (pesos/kg)	0.57	0.04	0.17	0.27	0.04	0.29
Number of samples	87	23	50	30	17	207

Sources: Compiled from Camargo Barros et al. (2003); Costales et al. (2003); Mehta et al. (2003); Poapongsakorn et al. (2003); Sharma et al. (2003).

Table 6.12 Environmental expenditures per unit output from milk production, 2003

Variable	Farm size/type				
	Small	Medium	Large	Commercial/ periurban	Total
	1–3 head	5–10 head	More than 10 head	More than 10 head	
India					
North zone expenditure (rupees/liter)	0.465	0.423	0.339	0.238	0.339
Number of samples	100	68	58	34	260
West zone expenditure (rupees/liter)	0.525	0.469	0.399	0.396	0.462
Number of samples	100	80	50	30	260
	1–20 head	21–50 head	More than 50 head		
Thailand					
Expenditure (baht/kg)	0.017	0.563	0.104		0.290
Number of samples	35	38	19		92
	50 or fewer head	51–70 head	More than 70 head		
Brazil					
Expenditure (reals/liter)	0.009	0.0095	0.008		
Number of samples	64	31	65		160

Sources: Compiled from Camargo Barros et al. (2003); Costales et al. (2003); Mehta et al. (2003); Poapongsakorn et al. (2003); Sharma et al. (2003).

in Table 6.11. Small-scale producers internalize more per kilogram than larger producers, except in the Philippines, where larger independent producers expend more per kilogram than medium independent producers but less than small producers. Larger producers in Thailand expend considerably less per kilogram than smaller producers.

Table 6.12 shows the environmental costs associated with dairy production. In India, the internalization costs per kilogram decline with increasing size. They are relatively constant in Brazil. Small producers spend very little on environmentally related costs in Thailand, but medium producers spend five times more than large producers.

Why Some Farms Spend More per Unit of Output on Environmental Mitigation

The results just reported showed that systematic differences in practices between large and small farms have led to persistent differences across the size distribution of farms in the internalization of negative environmental externalities per kilogram of livestock output. Across countries and commodities, small-scale farms, including contract farms, have higher environmental mitigation "expenditures" per unit of output than large-scale farms. The four exceptions are for poultry farmers in Thailand, where large-scale operations in the densely settled

Table 6.13   Determinants of farm expenditure on mitigation of environmental externalities from swine production, Thailand, 2002–03

Dependent variable: encovm (cost of environment abatement including manure sale)	Coefficient (N = 174)	Standard error	t-value
herd (number of total pigs in farm)	8.01	1.78	4.49***
agehh (age of farm owner in years)	−198.63	279.22	−0.71
female (1 if female farm owner)	−5,702.81	5,467.05	−1.04
yearedu (education of farm owner in years)	−641.59	700.07	−0.92
socstat (1 if social farm owner has status in a community, for example, holds a position in a local administration office)	5,745.73	6,126.14	0.94
distvil (distance to nearest village in km)	389.84	869.53	0.45
distriv (distance to nearest river in km)	627.36	265.53	2.36**
density (pig density in a radius of 1 km)	0.78	0.45	1.72*
flies (1 if farm owner is making an effort to reduce the number of flies)	9,562.00	5,270.01	1.81*
yearfarm (years of farm-breeding swine)	−79.63	313.95	−0.25
ltcredit (share of long-term credit to total borrowing)	11.28	52.38	0.22
crop (1 if farm owner receives income from crops)	563.47	7,852.58	0.07
fish (1 if farm owner receives income from fish)	2,476.70	10,250.09	0.24
nonfarm (1 if farm owner receives nonfarm income)	1,798.92	7,480.18	0.24
subs (environmental subsidy)	9,639.17	5,194.61	1.86*
chachern (1 if Chachernsoa Province)	273.80	5,545.07	0.05
cons (a constant)	−1,899.47	17,775.78	−0.11
Log likelihood = −1717.24			
Likelihood ratio $\chi^2(16)$ = 52.08			

Source of basic data: TDRI–FAO/LEAD (2002).
Notes: Costs of environmental abatement include variable cost (labor and others) and fixed cost (amortized value). Fixed inputs are water tanks, water treatment ponds, biogas ponds, water pipes. *, **, and *** indicate significance at the 10, 5, and 1 percent levels, respectively.

Table 6.14   Determinants of farm expenditure on mitigation of environmental externalities from broiler production, Thailand, 2003

Dependent variable: expenditure on environment cost (baht)	Coefficient (N = 132)	Standard error	t-value
Farm size (number of chickens per year)	0.259	0.055	4.73***
Female dummy	519.798	1,514.605	0.34
Log of age	–2,573.813	3,041.19	–0.85
Maximum years of education (operator or spouse)	–69.183	202.835	–0.34
Operator's years of experience	–480.574	264.285	–1.82*
Dummy if farm had been in family before the operator took charge	474.187	1,950.239	0.24
Distance to village	238.756	148.009	1.61
Distance to river	–220.79	278.199	–0.79
Broiler density in 1 km radius	–0.019	0.004	–4.78***
Dummy if farm has fish pond	615.142	1,550.058	0.40
Dummy if farm also has crops	34.267	1,857.096	0.02
Dummy if operator has nonfarm fish pond	3,561.87	1,695.88	2.10**
Constant	2,807.308	11,116.37	0.25
F-test (degrees of freedom = 12)	8.84***		

Source of basic data: TDRI–FAO/LEAD (2002).

Note: ** and *** indicate significance at the 5 and 1 percent levels, respectively.

export-certified zone spend more per unit than smaller farms; dairy farmers in Thailand, where the larger farms had more crop land than the smaller farms in the sample; and broiler and swine large-scale contract farmers in the Philippines, where manure and dead animal disposal practices were stipulated in the contract and where chicken manure is salable as fertilizer.

The next step is to determine factors associated with farms having to expend money to mitigate environmental externalities. To investigate this issue, we regress environmental mitigation expenditures on a series of exogenous household and farm characteristics. The regressions show that higher "expenditures" on the mitigation of environmental externalities are typically positively associated with raising a larger number of animals (except for broiler farms in India), being a mixed livestock-crop farmer (as opposed to a farmer operating a specialized livestock enterprise), being relatively far from

the nearest population cluster (except for independent and large broiler farms in the Philippines), being in a zone where there is active market demand for manure for use on crops (no surprise), and being more highly educated (Thai dairy farms only) (Tables 6.13–6.16; for the other nine regressions explaining determinants of environmental mitigation costs not presented here, see Camargo Barros et al. 2003; Mehta et al. 2003; Poapongsakorn et al. 2003; and Sharma et al. 2003). Greater concentrations of animals per farm area were negatively associated with environmental mitigation efforts in both Thailand (except for swine) and Brazil, where intensive operations are prevalent within the sample areas for this study.

Smaller farms are more likely to engage in mixed livestock-crop farming that is less specialized in livestock production than on large-scale commercial livestock farms. Not surprisingly, they tend to make a proportionately greater effort than large-scale operators

Table 6.15 Determinants of farm expenditure on mitigation of environmental externalities from broiler production, Philippines, 2002–03

Explanatory variable	Pooled sample (N = 116)	All contracts (N = 62)	All independents (N = 54)	All large farms (N = 54)	All small farms (N = 62)
Education of household head (yrs)	n.s.	n.s.	n.s.	n.s.	n.s.
Age of household head (yrs)	n.s.	n.s.	n.s.	n.s.	n.s.
Able to sell manure in last 2 months (dummy)	9,493.83 (1,865.97)	9,026.60 (3,016.00)	7,993.52 (2,000.51)	11,336.28 (3,845.92)	6,778.49 (1,967.98)
Class of land is agricultural (dummy)	n.s.	n.s.	−3,439.24 (1,898.37)	n.s.	n.s.
Land includes cropland (dummy)	4,079.71 (1,427.42)	n.s.	6,538.81 (1,827.28)	n.s.	4,280.35 (1,800.05)
Land is within the Laguna Lake Development Authority region (dummy)	n.s.	n.s.	3,888.58 (1,745.68)	5,852.79 (2,649.06)	n.s.
Wage rate (pesos/hour)	n.s.	n.s.	−418.02	n.s.	305.57
Wage rate (pesos/hour)			(189.59)		(160.68)
Number of mortalities in last batch	n.s.	n.s.	n.s.	n.s.	14.84 (6.92)
Distance to nearest residential community (km)	n.s.	n.s.	−1,786.47 (900.90)	−1,545.43 (888.16)	
Number of day-old chicks loaded in previous batch	0.26 (0.07)	n.s.	0.38 (0.06)	0.24 (0.09)	n.s.
Constant	−10,538.54 (4,327.64)	n.s.	−1,3061.17 (4,947.11)	n.s.	−7,810.60 (4,298.54)
Log pseudo-likelihood	−953.77	−561.40	−378.05	−501.31	−438.84
Likelihood ratio $\chi^2(10)$	74.06	26.46	69.09	37.03	44.65
σ	6,337.27	6,217.27	4,455.08	6,446.93	4,684.86

Source of basic data: UPLB-IFPRI (2003).

Notes: Numbers in parentheses are standard errors. n.s., statistically insignificant at the 10 percent level.

to spread manure as fertilizer so as to reduce their input costs on crop activities requiring nutrients. The ratio of animals kept on smallholder farms to the area of crops cultivated is smaller than on large farms, and the marginal utility of income from manure sales is undoubtedly higher. For example, Costales et al. (2003), Mehta et al. (2003), and Poapongsakorn et al. (2003) showed that having access to cropland for spreading manure distinguishes environmental mitigation behavior among sample farms belonging to the small farm substratum, but does not explain differences in mitigation behavior among large farms. Clearly, these results are also consistent with the mass balance calculations, which emphasize the more sustainable behavior of farmers who are mixed crop-livestock farmers.

Costales et al. (2003) also suggested an interesting scale difference with respect to location: large broiler farms closer to communities with large populations make less of an effort toward environmental mitigation. This is probably due to the fact that large broiler farms in the Philippines sample

Table 6.16   Determinants of farm expenditure on mitigation of environmental externalities from swine production, Philippines, 2002–03

Explanatory variable	Pooled sample (N = 203)	All independents (N = 152)	All contracts (N = 51)	All large farms (N = 95)	All small farms (N = 108)
Education of household head (yrs)	n.s.	n.s.	n.s.	n.s.	n.s.
Age of household head (yrs)	n.s.	n.s.	n.s.	n.s.	n.s.
Able to sell manure in last 2 months (dummy)	n.s.	n.s.	3,565.39 (1,187.08)	n.s.	2,037.97 (496.73)
Class of land is agricultural (dummy)	1,881.44 (1,084.49)	n.s.	1,146.74 (608.56)	n.s.	367.99 (193.68)
Land includes cropland (dummy)	857.10 (400.86)	1,082.47 (501.10)	n.s.	n.s.	347.40 (124.59)
Land is within Laguna Lake Development Authority region (dummy)	2,455.16 (1,167.16)	n.s.	n.s.	n.s.	n.s.
Wage rate (pesos/hour)	n.s.	n.s.	249.86 (106.36)	n.s.	n.s.
Number of mortalities in last cycle	n.s.	n.s.	18.44 (5.45)	23.25 (13.33)	n.s.
Distance to nearest residential community (km)	25.17 (12.89)	n.s.	10.99 (4.59)	n.s.	3.01 (1.69)
Land has connection to piped water		n.s.	1,675.25 (581.65)	n.s.	404.60 (245.16)
Number of swine in inventory	n.s.	n.s.	n.s.	n.s.	10.23 (5.69)
Region dummy (Bukidnon = 1)	n.s.	n.s.	n.s.	n.s.	n.s.
Constant	−15,116.67 (8,576.57)	n.s.	−6,569.59 (2,254.38)	−20,075.05 (11,954.5)	n.s.
Log pseudo-likelihood	−1,633.52	−1,240.36	−332.44	−934.84	−567.53
Likelihood ratio χ²(12)	37.74	47.36	45.12	19.96	30.98
σ	9,323.48	9,803.05	1,572.84	11,891.07	903.69

Source of basic data: UPLB-IFPRI (2003).
Notes: Numbers in parentheses are standard errors. n.s., statistically insignificant at the 10 percent level.

that are concentrated in periurban areas have the potential for spreading manure on their own farms.

Another determinant of environmental mitigation that seems to matter is contract farming. In the case of broiler farms in India, scale differences in environmental mitigation behavior tend to disappear, because both large and small contractors tend to have higher levels of measured environmental mitigation than independent farmers and not very different internalization across different sizes of contract farms. As mentioned earlier, the conditions for disposal of manure and dead animals were stipulated in the contract, adding some uniformity to procedures, along with a degree of environmental friendliness.

CHAPTER 7

Empirical Results

The present chapter synthesizes the key results of econometric models that were estimated separately for each commodity in each country. This involves 13 separate models: Indian broilers, eggs, and milk; Philippine swine and broilers; and broilers, eggs, milk, and swine in both Brazil and Thailand. Then separate runs were done in some cases for large and small farms within commodity/country categories and for contract farmers.

Separation of estimation by country and commodity is necessary because the farm-specific relative efficiencies computed by the Battese-Coelli (1995) model are computed relative to a frontier defined by the data in a given regression. While such farm-specific measures are comparable to each other within a given regression, they cannot be compared across separate regressions.[1] However, because the individual farm-specific measures are by construction independently and identically distributed random variables within a given run, the means of these farm-specific regressions of equal weight within a given regression are comparable to each other statistically.[2] For example, we can compare the mean relative efficiency of the bottom tercile of farms with respect to income to that of the top tercile in a given regression.

Comparison of Average Profit Efficiency across Farm Classes

Chapter 5 showed that small-scale farmers across the country studies typically earned higher profits per unit than large-scale farmers if family labor was not costed, which suggests that they at least have a chance to survive. It was also clear that costing family labor at market rates reduced this profit advantage for the smaller farmers, frequently eliminating it. If smallholders are not able to sustain a rate of productivity growth equal or greater than that of large farms under these conditions, they are likely to disappear.

The previous chapters and the first section of this chapter yield the data and methodology necessary to formally test Hypotheses 2 and 6 from Chapter 1, respectively: *Small-scale producers are more efficient users of farm resources to secure profits than are large-scale producers, other things equal* and *contract farmers are more profit-efficient than independent farmers at comparable scales of operation.*

---

[1]Battese, Rao, and O'Donnell (2004) illustrate a metafrontier approach that in theory could handle estimation of all cases simultaneously, or at least within commodity categories, thus permitting efficiency comparisons across countries. However, this ambitious task is left to future work.

[2]By equal weight we mean the same number of observations underlying each mean of a subsample group, as in, say, terciles. For example, we can compare the mean relative efficiency of the bottom tercile of farms with respect to income to the top tercile in a given regression.

Methodology and models are discussed in Chapter 3; the farm-specific results on relative efficiency in this section are estimates of a combination of technical and allocative efficiency, referred to here as "relative profit efficiency," measured on a scale of 0 (lowest) to 100 percent (highest, at the frontier). Farm-specific scores from a given regression are averaged over farms in the class in question to obtain an estimate of relative profit efficiency for that class of farm.[3]

An overall summary of results is discussed at the end of this chapter and shown in Table 7.15, but individual results will be discussed first. Table 7.1 shows a diversity of results for broilers, and Table 7.2 shows the case for layers. In the Philippines, larger broiler contract farmers have a slightly higher mean profit efficiency (73 percent) than smallholder contract farmers (56 percent), and the difference in the means is statistically significant (with s.e. = 0.03). But both large and small contract farmers have much higher average profit efficiency than either large or small independents. Larger independents (including independent forward-price contractors who are otherwise comparable to independents except for having a steady forward outlet) have greater relative profit efficiency on average than smallholders, and this difference is statistically significant at the 10 percent level (with s.e. = 0.04).

A graphical representation of the distribution of farm-specific relative efficiencies for broilers in the Philippines is displayed through kernel density plots in Figure 7.1a and b. It appears that large farms are more concentrated at higher levels of efficiency than are small farms. Similarly, contract farms are more concentrated at higher levels of efficiency than are independent farms, and the efficiency scores across farms do not have the same distribution functions (at a 1 percent level of significance).[4] This is consistent with the view that contracting improves the profit efficiency of both large and small farms.

Virtually all sampled broiler farmers in Thailand were contract farmers of some sort, with both forward-price contracts and fee contracts, with a degree of harmonization of genetics and feed use in both cases. The pooled (price contract / fee contract / independent) results for Thai broilers are reported in ascending order of size of flock in Table 7.1. The largest farms are the most profit-efficient (87 percent of maximum) compared to the smallest (48 percent of maximum). The bump in the medium-sized production level (5,000–20,000 birds) is probably due to the prevalence of fee contract farmers in this size of operation, which is likely to increase technical and allocative profit efficiency (if not profits) over price contracts at the same scale.

Similarly, in the case of Indian broilers, large independent farms are more profit-efficient than small independent farms. This result is further supported by the kernel density plot of the individual farm efficiency scores (shown in Figure 7.2), where large farms are more concentrated at higher levels of efficiency than are small farms, generating significantly different distribution functions (at the 1 percent level of significance using the Kolmogorov-Smirnov test for equality of distribution).

[3]Family labor is treated as a farm-specific physical input to specific activities and is not otherwise costed in any of the runs in this chapter, unlike in Chapter 5, where the sensitivity of unit profit levels to this assumption is specifically tested for the Philippines and India cases. Logically, the approach in the present chapter makes little or no difference within farms of similar size, but, ceteris paribus, will make smaller farms tend to look more profit-efficient than larger ones. Intuitively, the willingness of small farmers to work hard for low returns makes them relatively more profit-efficient users of other resources, ceteris paribus, compared to corporations employing formal sector labor.

[4]Based on the Kolmogorov-Smirnov test for equality of distribution functions (StataCorp. 2006).

Table 7.1 Mean relative profit efficiency of broiler farms across farm sizes by country, 2002

| | Farm size (number of birds) | | | |
| | Small (10,000 or fewer birds) | | Large/commercial (more than 10,000 birds) | |
Country	Independent	Contract	Independent	Contract
Philippines	N = 30 35	N = 34 56	N = 31 45	N = 14 73
		45		64
Brazil		N = 34 76		N = 195 86
	Fewer than 10,000 birds		10,000 or more birds	
India	N = 93 45		N = 42 85	
	Contract (5,000 or fewer birds)	Contract (5,001–10,000 birds)	Contract (10,001–20,000 birds)	Contract (more than 20,000 birds)
Thailand	N = 74 49	N = 51 71	N = 27 88	N = 18 87

Sources: Authors' estimation using data from CEPEA-ESALQ/USP-FAO/LEAD (2002); IIM-IFPRI (2003); RISDC-IFPRI (2003); TDRI-FAO/LEAD (2002); UPLB-IFPRI (2003).
Note: Mean relative profit efficiency is reflected by the percentage of maximum profit efficiency for the country and commodity in question unless otherwise indicated.

Table 7.2 Mean relative profit efficiency of layer farms across farm sizes by country, 2002

| | Farm size (number of birds) | | |
Country	Small (10,000 or fewer birds)		Large/commercial (more than 10,000 birds)
Brazil	N = 6 58		N = 83 72
	Small (10,000 or fewer birds)	Medium (10,001–50,000 birds)	Large (more than 50,000 birds)
Thailand	N = 47 52	N = 31 55	N = 18 61
	Small (fewer than 10,000 birds)	Medium (10,000–50,000 birds)	Large (more than 50,000 birds)
India[a]	N = 81 39	N = 63 36	N = 14 62

Sources: Authors' estimation using data from CEPEA-ESALQ/USP-FAO/LEAD (2002); IIM-IFPRI (2003); RISDC-IFPRI (2003); TDRI-FAO/LEAD (2002).
Note: Mean relative profit efficiency is reflected by the percentage of maximum profit efficiency for the country and commodity in question unless otherwise indicated.
[a]Farm size in India was grouped into terciles based on the annualized volume of output (in 100 eggs).

Figure 7.1   Distribution of mean profit efficiency scores for broiler farms in the Philippines

(a) Small farms versus large farms

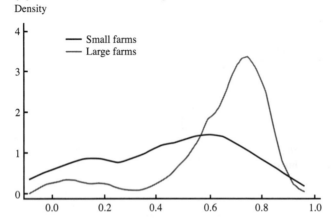

(b) Contract farms versus independent farms

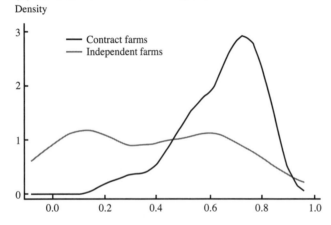

Source: Authors' calculations using UPLB-IFPRI (2003).

In Brazil (Tables 7.1 and 7.2), profit-efficiency levels are substantially higher for larger farms than for smaller ones for both broilers and layers; all broiler farms sampled in Brazil (in Table 7.1) are contract farms, while all layer farms sampled (in Table 7.2) are independent. The Indian layer farms in Table 7.2 are all independent. Here, the trend across increasing scales holds true between small and large farms, but in the case of small and medium farms, profit efficiency significantly decreases to the 15,000–50,000 egg per annum production level (at the 1 percent level of significance, with s.e.

= 0.04; also shown by the distribution of mean efficiency scores in Figure 7.3).

Insights about the relative profit efficiency of swine production are found in Table 7.3. In the Philippines, the mean efficiency of independent small-scale producers (77 percent) is higher than that of independent large-scale producers (67 percent). For contract farms, mean efficiency increases with size (Table 7.3). In Brazil (Table 7.3), medium and large producers are seen to be more efficient (71 percent and 72 percent of maximum profit, respectively) than small producers (61 percent of

Figure 7.2   Distribution of mean profit efficiency scores for broiler farms in India

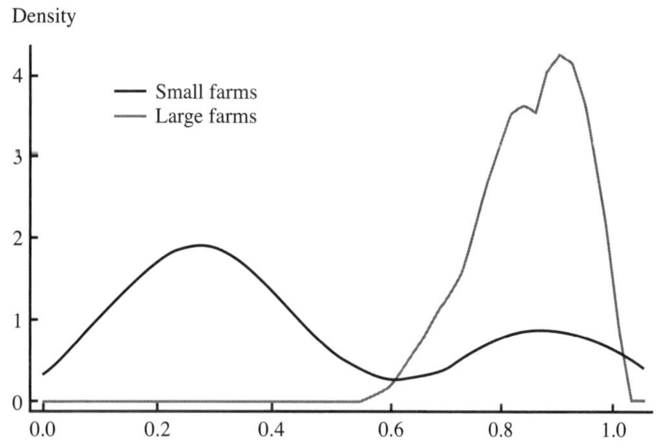

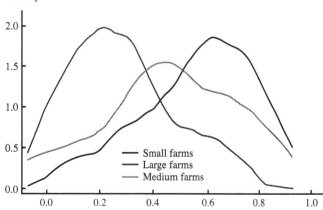

Source: Authors' calculations using RISDC-IFPRI (2003).

maximum), but the difference across farms is not significant (s.e. = 0.08). In addition, the mean efficiency scores between small and large swine farms have the same distribution functions (Figure 7.4). In Thailand (Table 7.3), mean profit efficiency in swine finishing increases marginally in going from backyard to small-scale commercial operations, then levels off with further increases in scale. For weaners, profit efficiency increases sharply in going from backyard to small-scale commercial operations, then to the very large-scale level, which is hardly surprising.

Unlike broiler, egg, and swine production, milk production is not usually thought to be subject to large technical economies of scale in production in developing country settings. Results from the three countries where milk production was studied are found in Table 7.4. Different patterns are observed in western India (Gujarat) and northern India (Haryana) (Table 7.4). In Gujarat, where the cooperative movement is very active and has done much to increase the profit efficiency of smallholders (rather as contract farming has for broiler producers in the Philippines), mean efficiency in-

Figure 7.3   Distribution of mean profit efficiency scores for layer farms in India

Source: Authors' calculations using RISDC-IFPRI (2003).

Table 7.3   Mean relative profit efficiency of swine farms across farm sizes by country, 2002

Country	Farm size (number of head)		
	Small (1–100 head)	Medium (101–1,000 head)	Large (more than 1,000 head)
Brazil	$N = 10$	$N = 59$	$N = 72$
	61	70	72
Thailand			
Weaners	$N = 7$	$N = 44$	$N = 3$
	85	89	94
Finishers	$N = 10$	$N = 52$	$N = 30$
	67	90	96

	Small/medium (1,000 or fewer head)		Large (more than 1,000 head)	
	Independent	Contract	Independent	Contract
Philippines	$N = 87$	$N = 23$	$N = 67$	$N = 30$
	77	64	67	75

Sources: Authors' estimation using data from CEPEA-ESALQ/USP-FAO/LEAD (2002); TDRI-FAO/LEAD (2002); UPLB-IFPRI (2003).

Note: Mean relative profit efficiency is reflected by the percentage of maximum profit efficiency for the country and commodity in question unless otherwise indicated.

Figure 7.4   Distribution of mean profit efficiency scores for swine farms in Brazil

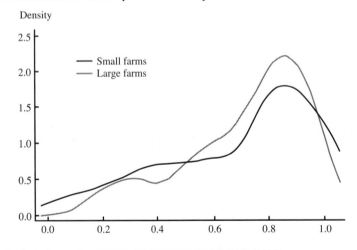

Source: Authors' calculations using CEPEA-ESALQ/USP-FAO/LEAD (2002).

creases with increasing farm production level. In Haryana, where larger-scale dairy farmers are more likely to be found than in western India, scale-related differences in mean efficiency are also observed.

Results for Brazilian milk production (Table 7.4) show decreases in mean profit efficiency up to 50–70 cows, although the results are not significant, and show increases thereafter. These results may be ex-

Table 7.4 Mean relative profit efficiency of dairy farms across farm sizes by country, 2002

Country	Farm size (liters of milk per day)			
	Fewer than 20 liters	**20–40 liters**	**41–80 liters**	**More than 80 liters**
India				
West	$N = 98$ 68	$N = 80$ 76	$N = 50$ 77	$N = 30$ 82
North	$N = 98$ 60	$N = 67$ 65	$N = 54$ 72	$N = 34$ 76
Pooled	$N = 196$ 64	$N = 147$ 71	$N = 104$ 74	$N = 64$ 79

	Farm size (number of head)		
	50 or fewer head	**51–70 head**	**More than 70 head**
Brazil	$N = 67$ 73	$N = 33$ 71	$N = 60$ 77

	20 or fewer head	**21–50 head**	**More than 50 head**
Thailand	$N = 32$ 56	$N = 33$ 60	$N = 19$ 73

Sources: Authors' estimation using data from CEPEA-ESALQ/USP-FAO/LEAD (2002); IIM-IFPRI (2003); RISDC-IFPRI (2003); TDRI-FAO/LEAD (2002).
Note: Mean relative profit efficiency is reflected by the percentage of maximum profit efficiency for the country and commodity in question unless otherwise indicated.

plained by the profit-efficiency advantage to smallholders of not costing family labor, which is less with increasing scale. It seems likely that not costing family labor has a lesser impact on relative profit efficiency as scale increases; furthermore, the largest scales may bring in new technologies and practices. On Thai dairy farms (Table 7.4), mean profit efficiency increases with scale, as in the case of Indian dairy farms. The difference in mean profit efficiencies between small farms and large ones is significant (s.e. = 0.04), but not that between small and medium farms or between medium and large farms, as shown in the distribution of their efficiency scores (Figure 7.5).

The overall results from comparing relative profit efficiency across countries, commodities, and degrees of vertical integration are mixed. Small dairy farms are not neces-

sarily less efficient at securing profits, especially once they reach the level of 20 cows per farm. Larger independent farms in the sample were clearly more profit efficient than independent smallholders for broilers and layers in all four countries. For swine, all farms operated at comparable levels of relative profit efficiency.

In almost all cases where contract farming was observed, the mean relative profit efficiency of contract farmers was significantly higher than that of independents at all levels of scale. Contract farming appears to significantly improve relative profit efficiency, more so for smallholders than for large farmers (except for small contract swine farmers in the Philippines), but significantly for both.

Hypothesis 2 (that small farms are more profit-efficient than large farms) is thus not

Figure 7.5   Distribution of mean profit efficiency scores for dairy farms in Thailand

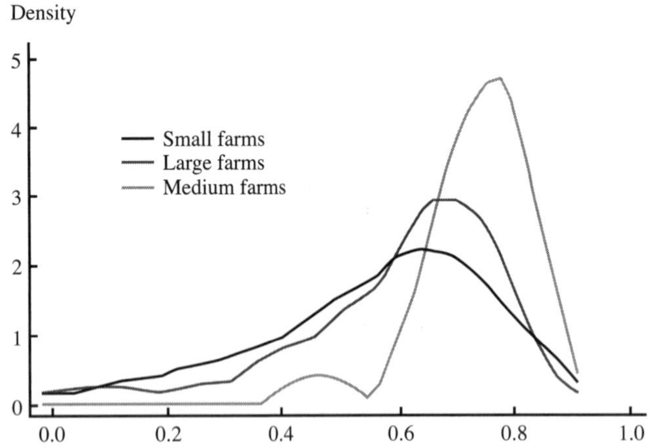

Source: Authors' calculations using TDRI-FAO/LEAD (2002).

well supported by the empirical analysis for broilers, layers, and milk, with the large farms exhibiting greater profit efficiency than small farms. However, Hypothesis 2 is well supported by the greater profit efficiency of independent small-scale hog fatteners vis-à-vis large-scale independent hog fatteners in the Philippines (statistically significant at the 1 percent level). The relative profit efficiency of smallholders increases when going from the backyard to the smallholder commercial model, but disappears fairly quickly with increasing size of operation, perhaps because the unit cost advantages of family labor stock become less important.

Hypothesis 6 (that contract farms are more profit-efficient than independent farms at comparable scales) is supported by the preceding results for poultry (both small-scale and large-scale operations) and swine (large-scale operations only) in the Philippines. The hypothesis is not supported by the results for small-scale swine producers in the Philippines, which was surprising to us. There is clearly more work to do in this area. Almost all dairy farming in the samples involved vertical coordination with cooperatives, which is also a form of contract farming. The relative profit efficiency for these dairy farms increases when going

from small to large/commercial operations. This is also true for broiler farms in Brazil and Thailand, where most of the farms sampled are engaged in contract farming. Thus, contract farming seems to work to improve efficiency (and thus competitiveness) by reducing transaction costs faced by independent small farms, as is investigated further later. The next section gives insights on which transaction costs and other factors explain why specific farms are profit-inefficient.

Explanation of Farm-Specific Profit Inefficiency in Terms of Differential Transaction and Environmental Mitigation Costs

The previous chapters and sections of this chapter yield the data and methodology necessary to formally test Hypotheses 4 and 5 from Chapter 1: *Uncompensated negative environmental externalities favor the nominal profit efficiency of large farms over small ones, ceteris paribus,* and *Profits of small-scale producers are more sensitive to "transaction costs" than are those of large-scale producers.*

Understanding why some farms are closer or farther from the stochastic profit frontier will clarify how differences in access to information, assets, subsidies, and the benefits of not paying for negative externalities associated with one's production affect the relative competitiveness of livestock farms. The measures of relative profit efficiency reported in the previous section are computed from the residuals of the frontier for the commodity and country concerned. An example of such a frontier is given for broiler farms in the Philippines in Table 7.5 (the pooled model). All coefficients reported in the tables are statistically different from zero at the 90 percent confidence level or better.[5] As discussed in Chapter 3, the coefficients in Tables 7.5–7.14 are maximum likelihood estimations of the influence of prices and farm fixed resources on maximum profits that a farm in the category concerned can make.

The previous section discussed the measure of farm-specific efficiency derived from the residuals (unbiased) of equation (3) in Chapter 3 of the stochastic profit frontier analysis and compared these across subgroups of farms within the same estimating equation.[6] The present section examines results from equation (4) of the model in Chapter 3. This regresses the residuals from equation (3) of the model (measures of the amount that a given farm falls below the frontier ideal) against a series of proxy variables for farm-specific transaction costs/policy distortions and our measure of farm-specific environmental mitigation (instrumented where appropriate).

Results for broiler producers in the Philippines are given in Table 7.5. In all the regressions except those for contract farmers, the key variable significantly explaining a reduction in relative inefficiency (that is, an increase in profit efficiency) within the subsample is the measure of environ-

mental mitigation. A negative coefficient means that the variable reduces inefficiency and explains why some farms are closer to the frontier, and this is a factor in both the large and small farm samples. Expending a higher degree of environmental mitigation effort is associated with being more profit-efficient for smallholders, large farmers, and independents, which are mostly large farms. However, the effect is not significant in the pooled regression involving all samples. The interpretation of this result is that those who pursue more environmentally sustainable practices in manure and dead animal disposal most likely have lower mortality rates and other efficiency gains relative to other farmers who are less careful.

Access to a telephone (typically a cell phone) and access to credit for feeds are significant factors distinguishing more or less relative profit efficiency among small farmers, but not among large ones (all of whom have at least cell phones and could buy feeds in bulk). Access to feed credit proxies, better access to assets, and access to phones proxy better access to information; both are significant transaction cost issues for smallholders.

For small farmers, higher education of the household head is associated with increased profit inefficiency, an unexpected result. The seemingly perverse effect of education on profit efficiency for smallholder farms may reflect the increased opportunity cost for more educated persons, particularly when the scale of operations of the broiler business is relatively small. In such cases, the attention of more educated people may be diverted to alternative economic undertakings (48 percent of the sample of small farmers are engaged in a secondary non-livestock occupation). Thus, the effect of education on livestock profit efficiency is not clear-cut.

---

[5]The null hypothesis that the coefficient is zero is rejected with less than a 5 percent chance of error using a two-tailed test.

[6]The residuals should be unbiased for the purposes of estimating farm-specific inefficiency and constructing the dependent variable of equation (4) in the model.

Table 7.5 Parameter estimates of stochastic profit frontier and determinants of profit inefficiency on Philippine broiler farms

Explanatory variable	Pooled (N = 109)	Independent (N = 55)	Contract (N = 54)	Smallholder (N = 64)	Large/commercial (N = 45)
Stochastic profit frontier					
Constant	21.99 (7.31)	32.65 (12.73)	5.03 (0.96)	20.11 (0.68)	−43.73 (1.09)
Broiler price	3.13 (1.67)	n.s.		1.47 (0.86)	14.59 (0.69)
Day-old-chick price	−7.07 (1.52)	−3.70 (1.27)		−4.79 (1.03)	n.s.
Feed price	n.s.	n.s.	1.38 (0.20)	4.94 (0.83)	n.s.
Wage rate	n.s.	n.s.	0.31 (0.16)	−2.24 (0.31)	0.48 (0.26)
Lowest interest rate	n.s.	n.s.		n.s.	n.s.
Buildings and equipment	n.s.	−3.82 (1.97)	1.99 (0.86)	−2.16 (0.74)	n.s.
Family labor	n.s.	n.s.	8.23 (0.97)	−11.23 (0.88)	3.88 (1.13)
Land	0.25 (0.06)	0.33 (0.08)	0.12 (0.06)	0.18 (0.02)	0.18 (0.08)
Family labor × wage	n.s.	n.s.	−3.49 (0.44)	5.22 (0.47)	1.81 (0.49)
Buildings and equipment × wage rate	n.s.	1.49 (0.72)	−0.78 (0.40)	1.13 (0.40)	n.s.
Buildings and equipment × interest rate	n.s.	n.s.	n.s.	n.s.	n.s.
Feed price × interest rate	n.s.	n.s.		−1.08 (0.15)	n.s.
Fee contract dummy	n.s.	n.s.			53.94 (0.96)
Log likelihood function	−155.08	−91.58	−42.25	−87.77	32.69
Likelihood ratio test (one-sided)	53.61	19.12	16.06	40.07	11.06
Explainers of inefficiencies					
Constant	n.s.	n.s.	n.s.	n.s.	n.s.
Age of decisionmaker	n.s.	n.s.	n.s.	n.s.	n.s.
Education of decisionmaker	n.s.	n.s.	−0.56 (0.26)	2.23 (0.86)	n.s.
Experience in poultry farming of decisionmaker	n.s.	n.s.	n.s.	n.s.	n.s.
Formal/informal training of decisionmaker	n.s.	n.s.	n.s.	n.s.	n.s.
Pick-up services provided by buyer	n.s.	n.s.	n.s.	n.s.	n.s.
Farm has phone service	n.s.	n.s.	−1.42 (0.65)	−3.55 (0.96)	n.s.
Environmental internalization	n.s.	n.s.	−0.09 (0.04)	−0.20 (0.04)	−0.17 (0.11)
Feeds credit (dummy)	n.s.	n.s.		−4.32 (0.57)	n.s.
γ	0.98	0.99	0.12	0.99	0.31

Source: Authors' estimation using UPLB-IFPRI (2003).
Notes: Numbers in parentheses are standard errors. n.s., statistically insignificant at the 10 percent level.

The results for Thai broiler producers in Table 7.6 suggest that the most efficient broiler farms are run by female decision-makers and with greater environmental mitigation efforts. Farms that tend to be clustered in areas with a higher density of chickens within a 1 km radius tend to be less profit-efficient. These farms are more likely to be located in traditional broiler farm areas, where conventional housing sheds (poorly ventilated sheds) are used, and this fact, correlated with location, may explain poorer production performance in terms of higher mortality rates and lower returns per bird.

In the case of the Indian broiler sample (Table 7.7), experience was a significant contributor to higher relative profit efficiency, particularly for large-scale farms, as expected. Furthermore, the more distant farms are from market outlets, the more relatively profit-inefficient they are, ceteris paribus.

On Brazilian layer farms (Table 7.8), regional and spatial differences are paramount to explaining relative profit efficiency. The most profit-efficient farms, ceteris paribus, tend to be those operated by older farmers and located in the south (São Paulo) and southeast (Paraná) regions. Characteristics that are associated with profit inefficiency are higher educational levels of decision-makers (probably due to additional nonfarm occupations, as in India and the Philippines), location in areas of higher concentrations of animals that are closer to big cities, and having access to information (which probably

Table 7.6 Parameter estimates of stochastic profit frontier and determinants of profit inefficiency on Thai broiler farms

Explanatory variable	Coefficient	Standard error
Stochastic profit frontier		
Constant	8.34	0.66
Log of output price	n.s.	
Log of feed price	n.s.	
Log of day-old-chick price	−0.27	0.15
Log of capital cost	0.53	0.06
Explainers of inefficiencies		
Delta (constant)	n.s.	
Farmer is female (dummy variable)	−7.34	1.86
Log of age of farmer	n.s.	
Log of farmer's years of education	n.s.	
Log of density of chickens (1 km radius)	0.53	0.21
Log of distance to nearest town (km)	n.s.	
Environmental internalization	−2.74	0.24
σ^2	39.09	8.30
γ	0.99	0.002
N	170	
Log likelihood function	−281.01	
LR test (one-sided)	225.97	

Source: Authors' estimation using TDRI-FAO/LEAD (2002).
Note: n.s., statistically insignificant at the 10 percent level.

Table 7.7   Parameter estimates of stochastic profit frontier and determinants of profit inefficiency on Indian broiler farms

Explanatory variable	Small[a] (N = 93)	Large[a] (N = 42)	Pooled[a] (N = 135)
Stochastic profit frontier			
Price of chicks	−0.77 (0.27)	−2.35 (1.00)	n.s.
Wage rate	n.s.	n.s.	n.s.
Price of feed	n.s.	−1.21 (0.53)	−1.58 (0.36)
Price of output (broiler)	3.59 (0.58)	5.59 (0.86)	4.64 (0.56)
Family labor	n.s.	2.11 (0.96)	n.s.
Value of capital stock	n.s.	−10.79 (0.73)	0.79 (0.39)
Wage rate × family labor	n.s.	n.s.	n.s.
Interest rate	n.s.	−54.29 (0.95)	2.79 (0.92)
Value of capital stock × interest rate	n.s.	8.22 (0.55)	−0.55 (0.29)
Constant	n.s.	146.30 (0.99)	−6.89 (1.17)
Explainers of inefficiencies			
Constant	n.s.	n.s.	−2.90 (1.21)
Age of the decisionmaker	n.s.	n.s.	n.s.
Information source (dummy variable = 1 for radio, TV, newspaper)	n.s.	n.s.	n.s.
Region (dummy, North = 1)	n.s.	n.s.	n.s.
Pollution abatement costs	n.s.	n.s.	n.s.
Has access to credit (dummy)	n.s.	n.s.	n.s.
Output market distance	0.64 (0.33)	n.s.	0.81 (0.16)
Experience	n.s.	−0.90 (0.33)	n.s.
Log likelihood function	−100.72	−36.39	−165.74
LR test (one-sided)	39.92	9.35	64.85
γ	0.94 (0.04)	0.51 (0.36)	0.15 (0.30)

Source: Authors' estimation using RISDC-IFPRI (2003).
Notes: Numbers in parentheses are standard errors. n.s., statistically insignificant at the 10 percent level.
[a]Contract farmers not included.

captures high transaction costs in gaining access to information). In addition, a higher degree of environmental mitigation effort is associated with being less profit-efficient, which could simply indicate that broiler farms are in remoter areas and tend to invest less in environmental mitigation.

For Indian layer farms (Table 7.9), being in the north and having an older operator tends to be associated with being more profit-efficient. Making greater relative efforts at mitigation of environmental externalities tends to be associated with being relatively less profit-efficient, as in the case of Brazilian broiler farms. If only the large-scale layer samples are considered (Table 7.9, second column), the relatively most profit-efficient farmers are those who are located in the north, are older, and are the best educated. In the case of these large commercial operations, higher education is more likely to be used to support the layer enterprise as opposed to commuting to a nonfarm job (contrary to what may well be the case for less industrialized smaller farms).

Table 7.8   Parameter estimates of stochastic profit frontier and determinants of profit inefficiency on Brazilian layer farms

Explanatory variable	Coefficient	Standard error
Stochastic profit frontier		
Constant	−0.317	0.106
Feed price	−0.400	0.026
Price of hired labor	n.s.	
Price of electricity	−0.180	0.006
Price of freight	−0.089	0.045
Land includes agricultural land	0.030	0.015
Family labor	1.039	0.007
Capital	0.111	0.017
Explainers of inefficiencies		
Constant	−4.395	0.763
Level of education of decisionmaker	0.919	0.332
Age of decisionmaker	−3.518	0.639
Length of time of decisionmaker in activity	n.s.	
Animal concentration on the farm	1.701	0.236
Environmental cost	1.087	0.578
Distance from the city	1.572	0.298
Information index	0.935	0.300
State dummy São Paulo	−2.221	0.899
State dummy Paraná	−7.458	1.686
σ^2	3.244	0.454
γ	0.999	0.000
Log likelihood function	−15.699	
LR test of the one-sided error	202.57	
N	89	

Source: Authors' estimation using CEPEA-ESALQ/USP-FAO/LEAD (2002).
Note: n.s., statistically insignificant at the 10 percent level.

Swine production in the Philippines has been analyzed, and several patterns emerge from the analysis of the determinants of relative profit inefficiency in Table 7.10. First, having access to information through media is an important driver of relative efficiency, particularly for smallholders who are making the key production decisions themselves. Second, farmers who sell to itinerant traders (*viajeros*) are typically less profit-efficient than farmers who have other marketing arrangements; the *viajeros* provide personalized service to backyard farmers, but at a price. Third, fee contractors are more profit-efficient for the overall sample than are independent and price contract farmers, other things equal.

Fourth, greater efforts to promote environmental mitigation are associated with increased profit efficiency for the overall sample and for large-scale farmers considered in isolation, but not for smallholders and independents as groups. A higher degree of environmental mitigation effort by smallholders suggests that they pay a price in terms of financial competitiveness. Some small-

Table 7.9   Parameter estimates of stochastic profit frontier and determinants of profit inefficiency on Indian layer farms

Explanatory variable	Small (N = 62)	Large (N = 96)	Pooled (N = 158)
Stochastic profit frontier			
Wage rate	80.75 (0.51)	1.89 (0.01)	0.91 (0.29)
Price of feed	–3.99 (1.21)	1.85 (0.008)	–1.94 (0.82)
Price of eggs	n.s.	–9.33 (0.02)	–5.29 (1.01)
Family labor	n.s.	–2.68 (0.03)	–2.89 (1.19)
Value of capital stock	n.s.	0.10 (0.001)	0.03 (0.01)
Slope of labor housing	0.04 (0.02)		0.04 (0.02)
Wage rate × family labor	n.s.	0.75 (0.01)	0.77 (0.32)
Wage rate × scale dummy	159.72 (0.95)		–0.41 (0.05)
Constant	12.94 (2.86)	–2.17 (0.04)	11.85 (1.59)
Explainers of inefficiencies			
Constant	n.s.	7.26 (1.15)	–3.78 (1.15)
Age of decisionmaker	–9.42 (4.90)	–41.59 (3.31)	–42.75 (5.79)
Region (dummy, North = 1)	–29.14 (4.29)	–23.51 (2.29)	–38.42 (2.59)
Pollution abatement costs	n.s.	9.54 (1.03)	11.66 (1.79)
Education of decisionmaker	12.02 (1.60)	–8.29 (1.90)	n.s.
Decisionmaker has access to credit (dummy)	n.s.	18.97 (1.46	n.s.
Information source (dummy variable = 1 for radio, TV, newspaper)	18.82 (5.50)	n.s.	7.60 (2.78)
Output market distance	n.s.	5.63 (0.77)	3.83 (1.02)
Log likelihood function	–74.65	–170.18	–272.14
LR test (one-sided)	139.79	148.35	243.14
γ	0.99 (0.0005)	0.996 (0.000)	0.99 (0.001)

Source: Authors' estimation using the RISDC-IFPRI (2003).
Notes: Numbers in parentheses are standard errors. n.s., statistically insignificant at the 10 percent level.

holders, unfortunately, can still get away with throwing manure over their back wall into the river instead of spreading it, and they make more money as a consequence.

Fifth, access to formal sector credit increases relative profit efficiency in the overall sample, among smallholders and independent farms, but a dummy for distinguishing access to credit for feeds in particular reverses some of the benefit of better credit (particularly for smallholders), suggesting that credit for longer-term capital investment is more important to relative profit efficiency for swine producers in the Philip-

pines than is short-term suppliers' credit for inputs.

The last column of Table 7.10 shows an interesting contrast for large-scale commercial pig farmers. Here, the only factors explaining better performance within the group are environmental cost mitigation (probably with an explanation similar to the case of Philippine and Thai broiler farms) and location (Bukidnon, the more profit-efficient zone, is a region of Northern Mindanao island far from Luzon, where the industrial swine sector is relocating as environmental pressures rise in Luzon). Another factor is

Table 7.10 Parameter estimates of stochastic profit frontier and determinants of profit inefficiency on Philippines swine farms

Explanatory variable	Pooled (N = 207)	Independent (N = 154)	Contract (N = 53)	Smallholder (N = 110)	Large/commercial (N = 97)
Stochastic profit frontier					
Constant	3.92 (1.67)	n.s.	n.s.	4.79 (1.75)	9.58 (1.08)
Price of hogs	0.63 (0.20)	n.s.	n.s.	1.30 (0.40)	1.61 (0.27)
Price of piglets	n.s.	n.s.		−0.41 (2.29)	n.s.
Price of weanlings	−0.19 (0.09)	n.s.	n.s.	n.s.	
Price of feed	−0.93 (0.20)	−0.98 (0.18)	n.s.	−0.46 (0.20)	−1.30 (0.29)
Contract dummy (1 if contract)	0.83 (0.38)			0.60 (0.30)	3.42 (0.96)
Wage rate	0.47 (0.17)	n.s.	n.s.	n.s.	0.11 (0.04)
Free board and lodging (dummy)	n.s.	n.s.	0.81 (0.39)	n.s.	n.s.
Lowest interest rate	n.s.	n.s.	n.s.	n.s.	n.s.
Value of inventory	0.17 (0.03)	0.53 (0.06)	n.s.	−0.68 (−0.04)	n.s.
Buildings and equipment	0.34 (0.13)	0.23 (0.13)	n.s.	n.s.	0.19 (0.06)
Family labor	n.s.	n.s.	n.s.	n.s.	n.s.
Farm size (in ha)	0.12 (0.04)	0.09 (0.04)	n.s.	n.s.	n.s.
Mortality rate	n.s.	n.s.	n.s.	0.03 (0.02)	n.s.
Farrow-wean (dummy)	1.77 (0.85)	4.16 (1.8)		6.28 (1.12)	8.70 (1.47)
Combined farrow-wean and farrow-finish	n.s.	n.s.		1.89 (1.02)	2.56 (0.62)
Family labor × wage rate	n.s.	n.s.	n.s.	n.s.	−0.13 (0.05)
Buildings and equipment × lowest interest rate	−0.18 (0.11)	n.s.	n.s.	n.s.	6.63 (1.37)
Explainers of inefficiencies					
Constant	−2.79 (1.43)	n.s.	n.s.	n.s.	n.s.
Age of decisionmaker	n.s.	n.s.	n.s.	n.s.	n.s.
Experience of decisionmaker in hog business	n.s.	−3.20 (0.32)	n.s.	−1.48 (0.79)	n.s.
Education of decisionmaker	n.s.	n.s.	n.s.	n.s.	n.s.
Formal/informal training of decisionmaker	n.s.	3.97 (1.26)	n.s.	3.89 (1.48)	n.s.
Feeds credit (dummy)	n.s.	n.s.	n.s.	2.86 (1.10)	n.s.
Capital credit (dummy)	−3.36 (0.82)	−2.21 (0.85)	n.s.	−2.58 (1.12)	n.s.
Information	−6.94 (1.42)	n.s.	n.s.	−5.30 (1.70)	n.s.
Buyer is a *viajero*	8.03 (1.55)	7.42 (1.47)		3.09 (1.10)	n.s.
Environmental cost	−0.17 (0.09)	0.36 (0.10)	n.s.	0.61 (0.21)	−0.11 (0.03)
Fee contract (dummy)	−6.25 (1.89)		n.s.		−0.86 (0.46)
Region (dummy; Bukidnon = 1)	−2.26 (0.85).	n.s.	n.s.	n.s.	−0.68 (0.20)
Log likelihood function	−263.32	−171.24	−49.45	−107.98	−101.17
LR test (one-sided)	50.07	60.24	14.89	72.42	27.99
γ	0.88	0.97	0.68	0.95	0.11

Source: Authors' estimation using UPLB-IFPRI (2003).

Notes: Numbers in parentheses are standard errors. n.s., statistically insignificant at the 10 percent level.

that vertically integrated operations (engaged in fixed-fee contractual arrangements) are more profit-efficient, even in the case of larger-scale operations. The proxies for access to information (such as education, age, experience, formal/informal training, and various sources of information, such as local media and cellular phones) and assets (feeds and capital credit dummies) are not significant for the regression with larger-scale farms and contract farms. This is not surprising, probably because all these larger farms and contract farms have solved their access issues, unlike the smaller farms and independent farms, where there is more variation. Thus, access to various sources of information, such as local media and cellular phones, was significant when large and small farms were pooled, but not for large farms alone.

Brazilian swine farms (Table 7.11) showed interesting results. Unlike dairy and layer farms, the profit-efficient swine farms were headed by more educated decision-makers who had been in the swine business longer and by those located in the south (Paraná) and center-west (Mato Grosso, Mato Grosso do Sul, and Goiás) states. The location effect could be explained by the fact that the swine industry has already begun to concentrate geographically in remoter areas of the center-west, which is also true in the case of broiler farms, but not to a greater extent than are layer farms, which are more concentrated in the southern and southeastern states. Farm characteristics that are associated with profit inefficiency are having older operators who have been managing the farm for a longer time. Age and experience seem to be related to a lack of enthusiasm to adopt new management skills and technologies. Another proxy for access to information is the information index that turned out to be associated with profit inefficiency. It is possible that the more informed the farmers are, the less dedicated they are to farm activities, probably because the contractor or integrator and not

the farmer is in charge of most of the technical and economic decisions on the farm.

The results for dairy farms are shown in Tables 7.12–7.14. As in several other cases described earlier, Table 7.12 shows that for Indian dairy farmers, hypothesized greater access to information (proxied by age, education, and access to different media sources) as an explainer of relative profit efficiency has a significant positive effect on the profit efficiency (that is, negative effects on inefficiency) of the small and large and commercial farm sample, but more so for small farms. Also, farms located farther from the market are less efficient, ceteris paribus. As for the medium-sized farms producing more than 20 but less than 80 liters of milk per day (with 4–10 milk cows), access to information has no significant effect on relative profit efficiency, which means that differential access to information did not distinguish medium-sized farms from each other.

For medium-sized and both large and commercial dairy farms in India, environmental mitigation cost had no significant effect on profit efficiency. It had a significant effect only on small-sized farms, where it shows up as increasing relative profit efficiency. The truth is probably that waste disposal is less of a problem in the Asian dairy context than in any of the other cases described earlier, because cow manure is actively sought as a soil amendment and even for fuel in some cases.

Thai dairy farms exhibit the same pattern as the larger-scale Indian dairy farms, and in fact most of the dairy farms in the Thai sample are close in size to the larger Indian dairy farms; for example, medium-scale dairy farms in Thailand (41 percent of the total dairy sample) have the same average number of in-milk cows (15) as the commercial dairy farms in India; Thai small-scale dairy farms (34 percent of the total dairy sample) have the same average number of in-milk cows (6) as the large-scale dairy farms in India. Differential ac-

Table 7.11   Parameter estimates of stochastic profit frontier and determinants of profit inefficiency on Brazilian swine farms

Explanatory variable	Coefficient	Standard error
Stochastic profit frontier		
Constant	0.677	0.259
Price of feed	−0.520	0.894
Price of hired labor	n.s.	
Price of output	0.490	0.156
Dummy complete cycle (1 = yes; 0 = no)	−0.163	0.043
Dummy integrate (1 = yes; 0 = no)	0.231	0.068
Land includes agricultural land	n.s.	
Family labor	n.s.	
Capital	n.s.	
Explainers of inefficiencies		
Constant	−0.897	5.335
Length of time of decisionmaker on this farm	1.413	0.570
Level of education of decisionmaker	−0.473	0.290
Age of decisionmaker	1.753	0.930
Length of time of decisionmaker in activity	−1.182	0.460
Animal concentration on the farm	n.s.	
Environmental cost	n.s.	
Distance from the city	n.s.	
Information index	0.300	0.131
State dummy Santa Catarina	−3.055	1.175
State dummy Paraná	n.s.	
State dummy Mato Grosso do Sul	−2.303	1.130
State dummy Mato Grosso	−3.757	1.725
State dummy Goiás	−3.018	1.675
State dummy Rio Grande de Sul	n.s.	
σ^2	1.067	0.462
γ	0.996	0.003
Log likelihood ratio	−25.16	
LR test	123.72	
N	141	

Source: Authors' estimation using CEPEA-ESALQ/USP-FAO/LEAD (2002).
Note: n.s., statistically insignificant at the 10 percent level.

cess to information (proxied by age and years of education), distance between farms and the nearest community and waterway, and environmental mitigation efforts did not help distinguish relative profit efficiency among Thai dairy farms. The relative profit efficiency of Brazilian dairy farms (Table 7.14) is driven positively by the degree of family involvement in the enterprise (the negative coefficient means that it is good for efficiency), the decisionmaker's being the owner and living on the farm, access to

Table 7.12   Parameter estimates of stochastic profit frontier and determinants of profit inefficiency on Indian dairy farms

Explanatory variable	Small farms (N = 196)	Medium farms (N = 147)	Large and commercial farms (N = 168)
Stochastic profit frontier			
Constant	5.91 (0.96)	5.63 (1.50)	2.43 (0.80)
Price of milk	n.s.	−0.45 (0.22)	0.42 (0.21)
Price of fodder	n.s.	n.s.	n.s.
Price of feed	n.s.	n.s.	−0.68 (0.20)
Family labor	n.s.	n.s.	n.s.
Wage rate	−0.93 (0.31)	n.s.	0.49 (0.19)
Family labor × wage rate	n.s.	n.s.	n.s.
Land	n.s.	n.s.	n.s.
Building and equipment	n.s.	0.16 (0.04)	0.07 (0.03)
Land × building and equipment	n.s.	n.s.	n.s.
Explainers of inefficiency			
Constant	−2.28 (1.22)	n.s.	n.s.
Age of decisionmaker	−1.59 (0.60)	n.s.	n.s.
Education of decisionmaker	−0.45 (0.09)	n.s.	n.s.
Distance of land from market	2.41 (0.82)	n.s.	n.s.
Access to information	−8.84 (0.67)	n.s.	−0.90 (0.49)
Access to credit	n.s.	n.s.	n.s.
Environmental cost	−1.27 (0.44)	n.s.	n.s.
Zone dummy	3.95 (0.90)	n.s.	1.29 (0.69)
γ	0.96	0.99	0.90
Log likelihood function	−198.42	−98.57	−105.52
LR test (one-sided error)	71.80	44.85	30.34

Source: Authors' estimation using IIM-IFPRI (2003).
Notes: Numbers in parentheses are standard errors. n.s., statistically insignificant at the 10 percent level.

Table 7.13   Parameter estimates of stochastic profit frontier and determinants of profit inefficiency on Thai dairy farms

Explanatory variable	Coefficient	Standard error
Stochastic profit frontier		
Constant	−10.627	4.721
Log of output price	7.580	1.911
Log of price of concentrate feed	n.s.	
Log of price of roughage	n.s.	
Log of capital cost	0.251	0.099
Log of farmland	0.207	0.105
Explainers of inefficiency		
Delta (constant)	n.s.	
Male operator (dummy variable)	n.s.	
Log of age of operator	n.s.	
Log of years of education (maximum years of operator's or spouse's education)	n.s.	
Log of distance of farm to community (km)	n.s.	
Log of distance of farm to waterway (km)	n.s.	
Environmental cost incurred (baht per head) (predicted value)	n.s.	
σ^2	n.s.	
γ	0.981	0.064
N	84	
Log likelihood function	−110.73	
Wald χ^2 Test (5)	36.66	
Number of restrictions	6	

Source: Authors' estimation using TDRI-FAO/LEAD (2002).
Note: n.s., statistically insignificant at the 10 percent level.

different media sources (information index), and location (as in the case of eggs and swine producers). Environmental mitigation efforts had no significant effect on profit efficiency for dairy farms in Brazil.

In sum, unlike the direct mass balance approach in Chapter 7, which clearly shows that large farms pollute more per unit of output, the indirect econometric approach of this chapter provides mixed results with regard to Hypothesis 4 (that uncompensated negative environmental externalities favor the relative profit efficiency of larger farms more than smaller ones). This is probably due to the facts that the environmental measures used in the econometric work are of

uneven quality across the country and commodity case studies and also that they are likely to be correlated—and in different ways across cases—with other factors that might be relevant to explaining profit efficiency, but were not included in the analysis.

Of the ten cases examined in Tables 7.5–7.14, the environmental mitigation variable was negative and statistically significant in four of the country/commodity regressions—for Philippine and Thai broilers, Philippine swine (large-scale farms only), and Indian milk (small farms only)—suggesting that greater environmental mitigation efforts decreased the distance from the profit frontier (that is, increased relative

Table 7.14   Parameter estimates of stochastic profit frontier and determinants of profit inefficiency on Brazilian dairy farms

Explanatory variable	Coefficient	Standard error
Stochastic profit frontier		
Constant	8.216	0.646
Humid feed price	−0.066	0.022
Dry feed price	n.s.	
Medicine price	−0.059	0.034
Genetic price	0.033	0.016
Hired labor price	n.s.	
Output price	0.554	0.111
Membership in a cooperative	n.s.	
Land includes agricultural land	n.s.	
Family labor	−1.340	0.091
Capital	−0.227	0.088
Explainers of inefficiencies		
Constant	−7.774	2.567
Decisionmaker is owner	−1.957	0.883
Decisionmaker is family member	−2.522	0.885
Decisionmaker has experience in dairy (years)	0.643	0.288
Decisionmaker lives on property	−1.659	0.653
Age of decisionmaker	1.239	0.703
Education of decisionmaker	n.s.	
Decisionmaker had training	1.638	0.632
Distance between the farm and the city (km)	n.s.	
Information index	−0.689	0.199
Environmental cost	n.s.	
State dummy Rio Grande del Sul	n.s.	
State dummy Santa Catarina	n.s.	
State dummy Paraná	n.s.	
State dummy São Paulo	−2.169	0.896
State dummy Minas Gerais	n.s.	
σ^2	1.342	0.185
γ	0.992	0.004
Log likelihood function	−38.826	
LR test of the one-sided error	229.32	
N	160	

Source: Authors' estimation using CEPEA-ESALQ/USP-FAO/LEAD (2002).

Note: n.s., statistically insignificant at the 10 percent level.

profit efficiency). The opposite was the case for Brazilian layer farms and large Indian layer farms. This is a case where the more environmentally friendly behavior of small and large farms may also be associated with other practices conducive to productivity but not adequately measured by the first equation in the frontier model.

The regressions for Indian layer farms (large farms only) and Brazilian layer farms both suggest that increased mitigation is associated with decreased profit efficiency. This suggests that increased expenditure on mitigation comes at a net financial loss relative to profits (as would probably be the case for small-scale and independent swine farms in the Philippines). Nevertheless, larger farmers are more likely than small-holders to be held accountable for pollution in all the countries studied. Therefore, internalization of the cost of pollution as a counterweight to the cost of mitigation may in some cases make mitigation expenditures better for profits and profit efficiency for larger farms than for smaller ones, ceteris paribus, as suggested by the ambiguous results for Philippine broiler producers in Table 7.5 and for Indian dairy farms in Table 7.12.

Thus, while the direct mass balance estimates of the previous chapter strongly suggest that large farms reap the benefits of not paying for their pollution to a greater extent than do small farms, the indirect econometric approach here has difficulty replicating this result across commodities and countries. The bottom line is probably that Hypothesis 4 is true, but not enough so to seriously affect the relative competitiveness of large and small farms to the extent of excluding the small from the market.

With regard to Hypothesis 5 (that the relative profit efficiency of small farms is more sensitive to transaction costs than is that of large farms), farm-specific transaction costs matter significantly in explaining relative profit efficiency in both the small and large farm samples, and this is a fairly consistent result across countries and com-

modities. On the whole, the econometric work does not support the "more" part of Hypothesis 5, except in the case of Philippine and Indian broiler farms and Indian dairy farms (particularly medium-sized farms with more than 20 but fewer than 70 milk cows). For smallholder producers, the main issues appear to be differential access to telephone service and the market information that goes along with this and access to formal sector or cheaper suppliers' credit for feed purchases. The same things matter for larger farms, but these farms appear on the whole to have dealt with access issues, so these variables do not serve well to explain interfarm differences in profit efficiency among larger farms.

Summary of Results for the Hypothesis Tests

What does all of this mean for smallholders? Table 7.15 shows a summary of results for the issues posed in Table 1.3.

Hypothesis 1: Small-scale producers have higher or equal profits per unit of output than do large producers. The results here were mixed, but for the most part independent smallholders did better on profits per unit of output than did large independent livestock farms in India and the Philippines, but not in Brazil and Thailand. The latter two countries and their country samples had larger farms than the first two, on average. It is probably not too much of a stretch to conclude that smallholder livestock producers were doing relatively worse in the more advanced agricultural economies of Brazil and Thailand. On the other hand, they seemed to be largely holding their own in India and the Philippines, at least for the time being, especially if family labor is not costed. If unit profits give a snapshot of current competitiveness, the next item yields insights as to where things will go in a market economy.

Table 7.15   Summary of empirical results

Hypothesis tested	Actual findings	Source
H1: Smallholders have profits per unit of output that are higher than or equal to those of large producers.	Independent smallholders have higher profits per unit than large farms: broilers in the Philippines and India, layers in India (w/o costing family labor) and Thailand, milk in India and Brazil (n.s.), and swine in the Philippines. Small contract farms in the Philippines did better than large ones in the case of broilers (n.s.) and swine. Large contract farms did better than small contract farms in the case of broilers in India (w/o costing family labor), Thailand, and Brazil (n.s.).	Tables 5.4–5.7
H2: Smallholders are more profit-efficient users of farm resources than large producers, other things equal.	In cases where significant numbers of small-scale and large-scale farms were observed, only one case has shown that smallholders have greater profit efficiency than large ones, and this is the case of independent swine farms in the Philippines.	Tables 7.1–7.4
H3: Small farmers expend a greater amount of effort/investment in abatement of negative environmental externalities than do large farmers.	Smallholders in all cases, except large dairy farms in Thailand, pay more per unit of output in mitigating environmental externalities than do large farmers.	Tables 6.10–6.12
H4: Uncompensated negative environmental externalities favor the nominal profit efficiency of large farms over that of small ones.	Large farms, particularly layer farms in India, are polluting more, but evidence is mixed: for large Indian layer farms, being environmentally friendly comes at a price and hurts competitiveness; for large Philippine broiler and swine farms, spending more on environmental mitigation gives them a profit edge by achieving lower mortality rates and other efficiency gains relative to other farmers who are not environmentally friendly.	Tables 7.5, 7.7, 7.9–7.10; Figures 6.4–6.7
H5: Profits of small-scale producers are more sensitive to transaction costs than are those of large-scale producers.	Profits of smallholder swine producers in the Philippines and dairy farmers in India were sensitive to transaction cost variables in terms of access to information.	Tables 7.10 and 7.12
H6: Contract farmers are more profit-efficient than independent farmers at comparable scales of operation.	In cases where significant numbers of contract and independent farms were observed, only small and large contract farms for broilers in the Philippines were more efficient than independent farms at comparable scales of operation; small independent swine farms were more profit-efficient than small contract swine farms.	Table 7.1 and Figure 7.1b; Table 7.3

Note: n.s., not significant.

Hypothesis 2: Small-scale producers are more efficient users of farm resources to secure profits than are large-scale producers, other things equal. Unfortunately for smallholders, this hypothesis was supported only for the Philippine swine case when considering independent farms, and even there the relative advantage of smallholders over large farms was small. The hypothesis was not supported for any of the contract farming cases, although smallholder contract farmers in the Philippines were more profit-efficient than independent smallholders producing broilers.

Hypothesis 3: Small farmers expend a greater amount of effort/investment in abatement of negative environmental externalities per unit of output than do large farmers. As shown in Chapter 6, the hypothesis was supported by all samples except those of larger-scale dairy producers in Thailand. Regardless of motive or end use, smallholders clearly "spent" more resources in terms of time and money in dealing with animal waste.

Hypothesis 4: Uncompensated negative environmental externalities favor the nominal profit efficiency of large farms over small ones, ceteris paribus. This hypothesis is supported by statistically significant results only in the stochastic profit frontiers for large farms producing broilers and swine in the Philippines. There may be significant spin-off effects of environmental mitigation efforts that our analysis cannot fully explore. For example, larger farms that spend more on environmental mitigation may recoup their investment by achieving lower mortality rates and other efficiency gains relative to other farmers who are not environmentally friendly. Further, large farms that spend a lot on environmental mitigation may also recoup their investment by selling animal waste as fertilizer.

Other things equal, the results also suggest that a large environmental mitigation effort was actually associated with *lower* relative profit efficiency for small swine farms in the Philippines and for large Indian layer farms. For these farms, being environmentally friendly comes at a price and hurts competitiveness vis-à-vis those who get away with polluting.

On the other hand, the mass balance results clearly show that small farms are in a much better position than large ones to absorb excess animal wastes on their land. If they in fact do so, as suggested by the higher mitigation expenditures of smallholders per unit of output, it could be inferred that they are polluting less than large farms and thus are internalizing a higher share of the negative externality from animal waste.

On balance, the evidence is not convincing that uncompensated pollution is giving large farms a decisive profit edge over small farms, although it seems likely that they are in fact polluting more per unit of output and that if environmental regulations were fully enforced in the sample countries, it would cut into unit profits at least as much and probably more for large farms than for small ones.

Hypothesis 5: Profits of small-scale producers are more sensitive to "transaction costs" than are those of large-scale producers. Although the relative profit efficiency of most farm groups investigated was sensitive to the transaction cost variables included, this was unique to smallholders in the case of India broilers and was actually contradicted in the runs for Indian layers, where these variables were much stronger for larger farms.

Hypothesis 6: Contract farmers are more profit-efficient than independent farmers at comparable scales of operation. Although only the Philippine case study had large enough samples of large and small contract farmers and large and small independent farmers to allow for a robust four-way comparison, this hypothesis was well supported by Philippine broiler farms but not by swine farms. Smallholder independent swine farms are more profit-efficient than small contract producers.

CHAPTER 8

Conclusions

This study has assessed the outlook for smallholders in livestock farming in a selection of fast-growing developing countries, given the growing share of output coming from larger farms, particularly in the case of pig and poultry products, which have accounted for roughly three quarters of the growth in developing country livestock production since 1980. The potential negative impact of increased livestock production on the environment was also assessed in terms of how it affects and is affected by the scaling-up of individual farm sizes. The overall objective was to explore the scope for policy measures to provide a more level playing field for smallholders if in fact the fast pace of scaling-up is unduly driven by factors other than nondistorting incentives, such as economies of scale in technology or management.

The empirical approach lay in distinguishing "genuine" economies of scale (such as those resulting from lower unit prices for feed bought in bulk and from higher feed conversion ratios) through the specification of a stochastic profit frontier model with inefficiency effects (Battese and Coelli 1995; Coelli, Rao, and Battese 1998) from "false" economies of scale (such as scale non-neutral subsidies, externalities, and transaction costs). The frontier model yields a view of the average relative profit efficiency of specific groups of producers. Coupled with the actual profit performances of the groups in question, this approach indicates which groups are likely to be more competitive over time. The inefficiency term yields insights as to how much the frontier model results are driven by factors that are distortionary and that, in principle, should and can be influenced by policy action.

All the country studies confirmed that livestock production was both growing and concentrating rapidly, with more and more animals being kept per square kilometer. Production in the 1980s and 1990s tended to concentrate around capital cities and other major demand centers. Concentration over the period studied tended to be led by large-scale investments in monogastric livestock production, and there is evidence of significant environmental issues associated with this expansion. Since the late 1990s, in all study countries, expansion of larger farms has begun to move away from periurban areas to remoter areas where environmental issues, disease control, and feed supply are less problematic.

The country studies also indicated that smallholder output is continuing to grow in most cases, and at high rates in cases such as those of milk in India and swine in the Philippines. Smallholder dairy producers in Thailand have persevered largely due to high levels of policy support. For some commodities, such as broilers and eggs, smallholders in all four countries have been rapidly losing market share. Although there have been smallholder exits from poultry activities in all countries, some former smallholders from the early 1990s are now large farmers in all countries, sometimes migrating to new areas, such as the center-west in Brazil, or expanding in situ, as in southern India.

Profits and Relative Profit Efficiency: Who Does Best across Size Groups?

Comparative Unit Prices and Profitability by Class of Farm

The results here, detailed in Chapter 5 and the country studies, show that in several cases if family labor is not costed, independent smallholders made statistically significantly higher profits per unit than larger-scale farmers (7 out of 11 cases examined). Cases in point were slaughter hogs in the Philippines, broilers in India and the Philippines, eggs in India and Thailand, and milk in Brazil and India. Large farms had higher unit profits in the cases of swine, milk, and broilers in Thailand and in the cases of eggs and swine in Brazil. This finding disappears for contract farms, however, where in three out of five cases large contract farms did better per unit than small farms and there was no statistically significant difference between unit profits of large and small farms for the other two cases. The unit profit advantage of independent smallholders vis-à-vis large independents also largely vanishes in the Philippines if family labor is costed at market rates, and it disappears altogether in India, even though unit profits themselves remain positive. As argued previously in this report, costing family labor at full market rates is inappropriate, so these results should be thought of as a lower bound on smallholder competitiveness.

As noted in Chapter 3, the farm-specific average output prices collected varied significantly across farms. Large farms in the surveys received on average higher output prices per unit than small farms; the two exceptions in the country surveys were independent broiler farms in the Philippines and contract hog farms in the Philippines. Further research might well be directed to explaining this anomaly. As discussed previously, prices reflect a variety of supply and demand factors. On the supply side, larger farms that are more efficient may increase their competitiveness by discounting prices to bulk buyers.

On the demand side, buyer perceptions of higher quality and safety in relatively more branded products from larger farms command price premiums in the market.

It is difficult to know how much of buyer perception is due to true quality differences and how much is due to transactions costs, which, as shown in Chapter 7, has a disproportionate effect on smallholders. If a buyer is not sure that a product is safe or of high quality—neither one of which can be observed at sale—and the farmer cannot brand the output somehow, buyers will pay only the lowest common denominator price. Larger farms with higher and steadier volumes of output can build up a steady clientele and establish market trust and reputation more easily than can smallholders. The same is true of integrators.

The fact that smaller farms typically receive lower prices for their output and still manage to realize higher unit profits in some cases is only partly explained by not costing family labor, which, ceteris paribus, lowers the unit costs of smallholder producers more than those of large producers. The quality of supervision and care exerted by smallholders over their own stock is clearly superior to that in large enterprises in some cases as well. Given the higher level of output prices received by large farms and the lower prices per unit paid for purchased inputs such as feed, as shown in Chapter 5, the issue of increasing output prices is of particular importance to smallholders.

Two issues are critical here: how to improve the actual quality of smallholder livestock products and the separate issue of how smallholders can gain recognition in the marketplace for improved quality when their sales are infrequent and increasingly anonymous as market chains become longer. Insights on this come from the five cases of contract farming studied, involving slaughter hogs in the Philippines and Thailand and broilers in India, the Philippines, and Thailand.

There was less variation in prices received by large and small contract farmers vis-à-vis each other than was the case for large and small independents. There was also not much of a gap between prices received by large independents and those received by contract farmers, large or small. Scale-related differences in unit profits for contract farms are driven more by the cost side, unlike in the case of the independent farmers. It seems likely that the contract farming schemes for monogastric livestock serve in some sense to brand smallholder products, certifying the quality of output based on the certifiable quality of inputs provided by integrators.

Although further, more action-oriented research is required in this area, it seems clear from the present study that contract farming is a promising way to help small farms establish market trust and reputation and thus to maintain their competitiveness vis-à-vis large independent farms. Even without contracting and with costing of family labor, smallholders in some commodities in the Asian cases were able to achieve higher unit profits than larger farmers, suggesting that the latter will have some difficulty driving the former out of the market for some time to come. However, the outlook for small farms in Brazil is not very rosy according to the results here, nor is the two-cow dairy farm likely to be able to survive long in Thailand with further subsidy, unless producing milk is a hobby activity.

Furthermore, even the cases where smallholders had higher unit profits need to be kept in perspective. In the Philippines for example, the total average annual income from swine-raising for farms in the independent smallholder sample was US$309 per farm, whereas for the larger farm sample it was US$9,650 per farm. The latter does not include the profits of the largest farms, which belong directly to integrators and were not surveyed here. Even more pronounced gaps could be observed between the incomes of large and small farms in Brazil and Thailand. Thus, despite the better per-unit profit of smallholders, large farmers might still drive them out if

they are more efficient users of resources for production.

Comparative Profit Efficiency

Unlike profit per unit, which is a descriptive variable, relative profit efficiency per farm is an analytical result from a modeling effort. Chapter 7 computes and compares mean relative profit efficiencies across countries, commodities, scales of farm, and degree of vertical integration. Overall, the results support the view that small farms are less efficient at securing profits than large farms, even when family labor is not costed, and more so if it is.

Of the 12 country/commodity cases reported in Chapter 7 where sample size permitted testing small-scale versus large-scale farms without costing family labor, only one unambiguously showed a greater mean relative profit efficiency for smallholders. This was the case of independent swine farmers in the Philippines. When this case is re-run with profits reflecting costed family labor, the greater relative efficiency of smaller operations largely disappears. For the three cases involving independent farm samples tested, the large farm group did better on average, although the difference was not significant in two of these cases (layer and swine farms in Brazil). In all four of the cases where sample size permitted testing smallholder contracts versus larger scale contracts, the latter did significantly better.

The examples of swine in Brazil and Thailand, poultry in Thailand, broilers in India, and milk in India and Thailand show that the efficiency advantage of smallholders increases substantially when going from the smaller backyard producers to the smallholder commercial model, but relative efficiency is fairly static in the middle ground and rises again only with much larger sizes of operation. In part this is because the unit cost advantages of smallholders in the calculations arising from an uncosted stock of family labor become less important as output increases. More important, there appear to be especially significant profit-efficiency gains in going from finishing, say, 15–20 piglets a year, as in the backyard case, to finishing 150–200, as in the smallholder commercial case.[1] The smallholder commercial model, which is really a scaled-up and improved version of the backyard model, would seem to be a viable route in Asia for technology and institutional development targeted at improving smallholder livelihoods through swine, and it will probably require some form of vertical coordination to be successful over the medium term.

Milk production clearly suggests greater profit efficiency in the Asian sample countries when going from relatively small to large-scale operations (consistent with roughly 15–30 milking cows). All cases of milk production suggest that there are greater profit-efficiency gains for much larger farms. It is not clear from the results whether milk production is and will continue to be a smallholder activity in Asia. It seems likely that 1- to 2-cow dairy farms will grow in size to 10- to 20-cow farms in the Indian and Thai contexts over the medium run.

As shown in Chapter 7, vertical coordination such as fixed-fee contract farming and dairy cooperatives clearly improves the relative profit efficiency of smallholder farmers, even if unit profits were typically lower for contract farmers because capital costs and risks are shared by the integrator.[2] Contract farming works to improve efficiency (and thus competitiveness) by reducing transaction costs. The next section gives insights on which transaction costs and other factors explain why specific farms are profit-inefficient.

All of this abstracts from possible economies of scale in collection, processing, and distribution of products such as milk

and poultry, where scale economies on the marketing (not production) side may be major.[3] The future of smallholder livestock farming in developing countries will be largely driven over the medium to long run by the issue of whether collective action such as that of producer associations or cooperatives or contract farming schemes can reduce the transaction costs facing individual smallholders sufficiently that they can continue to compete with larger farms. The next section will look in detail at the factors that explain why specific farms were more or less profit-efficient.

Why Are Some Farms More Efficient at Making Profits than Others?

The same analytical approach that yielded the results on efficiency levels in the previous subsection also permits assessment of the determinants of relative profit efficiency across farms. In effect, we are simultaneously explaining why some farms are less profit-efficient than others in terms of cross-farm differences in environmental mitigation behavior, differences across farms in access to information and assets, and differences (if any) in access to policy subsidies.

The Role of Environmental Externalities

Two indirect approaches were used to get at scale differences in the capture of environmental externalities (in other words, why one farm incurs costs while another receives the benefits). The first approach measured the ability of a household to absorb the nutrients created by its livestock operations on the household land surrounding the operation by calculating individual household

---

[1]That is, going from the annual output of a single sow to that of 8 to 10 sows.

[2]The Philippine and Thai cases show that profits can actually be higher for contractors in situations of acute competition where even large-scale independents are having trouble remaining profitable.

[3]This assertion is not investigated by this study but is widely and probably correctly believed.

mass balances. The second approach constructed an index of the cost of mitigating potential environmental problems per farm, which was then used as a regressor in the inefficiency term of the stochastic profit frontier as an indirect measure of the farm-specific degree of internalization of environmental impact. It was not practical to perform more direct measures of environmental harm specifically attributable to the activities of each farm.

Mass Balances. The mass balance calculations were used as rough indicators of potential soil nutrient loading from livestock waste. The mass balance estimates for swine production in all the study countries showed a range of excess nitrogen and phosphorus balances across farms. For each of the three countries in which swine production was sampled, the larger producers showed a greater degree of nutrient loading and thus a need to specifically implement adequate disposal methods for manure. Each country had some households with sufficient land to assimilate the nitrogen produced. Smaller producers were much more likely to have positive balances from the operation of crop land that could potentially absorb the nutrients. Whether or not they use the manure this way is another matter; the Philippines survey reported that some smallholders simply dump swine waste on the slopes leading to creeks. Brazil, with its larger amount of land availability, had many households with positive balances in every size category. However, most of the sampled large-scale producers in the rest of the study countries had negative balances stemming from inadequate land on which to dispose of the manure generated from their livestock operations.

Nutrient balances for poultry showed wide variation between countries. The household surveys indicated that large producers are much more likely to have large negative mass balances, while smaller producers are more likely to have very small deficits in absorption or none at all. The more nega-

tive the mass balance, the greater the risk of nutrient loading. However, in some of the study countries large- and medium-scale producers were found to have small deficits (only faintly negative mass balances). In Thailand, most small- and medium-scale broiler producers had only mildly negative mass balances, although large-scale producers had more negative mass balances, and some of them were quite negative. Only India had small producers with nutrient balances of less than –10 metric tons, indicating a more uniform potential among smallholders for sustainability. However, India had no households with positive balances, which is indicative of the higher number of households producing on relatively small plots of land.

Relative to pig and poultry production, the mass balances for dairy production are comparatively in balance. Dairy production, by its nature, requires land. Thus, producers may have sufficient land to properly dispose of manure or may have access to nearby land. It is possible that small-scale producers in India or Thailand may not need sophisticated manure disposal mechanisms as long as the ratio of animals kept to land remains at the current average levels. On the other hand, the semiarid areas of these countries are largely deficient in soil organic content, and there cow manure is a vital soil additive.

Environmental Mitigation Effort. A second empirical approach was also used that gets around the issue of whether households actually exploited the potential resources they have for environmental mitigation. This yielded a farm-specific measure of environmental mitigation in terms of money units of mitigation effort per farm and per unit of output. This assumes that negative externalities within a country and commodity group are equal for each unit of output. Farmers who make an effort, financial or otherwise, to prevent problems by spreading manure or otherwise cleaning up or compensating their neighbors for problems created are, by definition, internalizing a portion of the negative externalities created per unit of output. If externalities are the same per unit of output, greater mitigation effort (measured in terms of money spent per unit of output) means greater internalization, other things equal.

All the country studies found that for broilers, smaller producers mitigated more per unit of output. However, this difference in mitigation effort by size was striking. Smaller producers scored on average more than five times more in environmental mitigation effort per unit of output than large farms in each country except Brazil. In the latter case, the difference was a factor of two. For Thailand, the mitigation efforts per unit for egg-producing households were greater than for broiler-producing households. In Brazil, the opposite was seen. In the Philippines, small contract broiler producers were found to have made a smaller effort on environmental mitigation than independents, but larger contactors made more of an effort than independents.

Small-scale swine producers also made a greater effort per unit of output than did larger producers. The exception was found in the Philippines, where larger independent producers expended more per liveweight kilogram of output than medium-scale independent producers, but less than small-scale producers. Larger producers in Thailand expended considerably less per unit than smaller producers. In India, dairy producers' environmental effort per liter of milk output declined with increasing size. Efforts were relatively constant across scales in Brazilian dairy. Small producers expended very little on environment-related costs in Thailand, but medium-sized producers expended five times more than large producers.

On the whole, the mass balance and mitigation expenditure approaches are consistent. As discussed in Chapter 6, smallholders make a significantly greater effort to mitigate negative environmental externalities than larger-scale farmers. However, the absolute orders of magnitude do not

suggest that this is a major explainer of scaling-up, although it may be one factor.

Chapter 6 also showed that results are mixed with regard to whether large farms are helped more by uncompensated negative externalities associated with their livestock production than is the case for small livestock farms. Differences in environmental mitigation across farms did not appear to be strong explainers of differences in relative profit efficiency.

The brunt of the evidence is that within the class of large-scale operations for swine and poultry, greater effort to mitigate environmental externalities seems to be associated with greater relative profit efficiency. This is clearest in the case of broiler and swine producers in the Philippines. Interestingly, these tend to be mostly independent operations that are transitioning toward more industrial production, at least compared with other samples studied. It is conceivable that more environmentally responsible behavior on the larger farms is correlated with the use of other best practices that, as a whole, boost agricultural productivity. The results for smallholders are more mixed, especially if contract and independent subsamples are considered together. Most fee (or wage) contractors have to follow a standard set of environmental practices as part of their contract, and they resemble larger-scale farmers in this respect more than other smallholders.

The environmental mitigation variable did not seem to have much influence on relative profit efficiency in the Brazilian and Thai dairy, Indian dairy (medium- and large-sized farms), India broiler, and Brazil swine samples. Egg producers in Brazil and India and smallholder swine producers in the Philippines that spent relatively more on environmental mitigation tended to have lower relative profit efficiency at the end of the day, other things equal. It is interesting to speculate whether these subsamples operated in conditions where it was relatively easier to ignore environmental issues or perhaps harder to follow environmentally sound practices because of land scarcity.

The Role of Access to Information and Assets

Farm-specific transaction costs seem to matter greatly in explaining relative profit efficiency across farms in most of the subsamples studied, large and small. This means that relatively greater difficulties in securing access to assets and information are prime explainers of differences in relative profit efficiency within the group in question. However, the differential impact across sizes of farm was different in different countries and for different commodities.

Milk production is one sector where farm-specific differences in transaction-cost proxy variables did little to explain differences across farms in relative profit efficiency. It is likely that transaction costs for milk production almost all occur in the marketing chain (where they may be very high in fact; witness the success of dairy-marketing co-ops worldwide) and not at the level of production. Feed is mostly forage (avoiding the high credit- and quality-related transaction costs packed into using concentrate feeds), and the timing of sales is a foregone conclusion, viz. daily.

Yet even if transaction costs occur in distribution and processing, they may be greater for integrators or dairy cooperatives dealing with smallholders than for those dealing with larger farmers. This is the conclusion of the Brazil country study as to why smallholders (in the Brazilian sense) are leaving milk production so quickly in Brazil: they are considered too costly to service by dairies that are free to not do so since the recent repeal of the cooperative law.

Transaction costs play a much more defining role in the production of monogastrics, where the timing of sale is more discretionary, the quality of inputs (genetics and feed concentrates) is critical to quality (requiring information), and 85 percent or the cost of production needs to be spent up front in feeder stock and feed. Broiler and egg sellers in the Indian samples, for example, had profit margins of less than 4 percent but could face up to 10 percent daily fluctuations in prices. Keeping a broiler on feed for an extra two days beyond the usual 42-day cycle would wipe out the profit margin. Philippine smallholder hog finishers who could not convince buyers that they had animals with low percentages of backfat and no off-flavors of meat were paid up to 10 percent less per kilogram liveweight for hogs compared to contract farmers and large farms. For those cases where transaction cost variables matter most to smallholder producers, the main issues appear to be access to various sources of market information and access to credit.

With regard to Hypothesis 5 (*Profits of small-scale producers are more sensitive to transaction costs than are those of larger producers*), the nuanced conclusion is that transaction costs matter greatly to the relative efficiency of smallholders, but they seem to matter to large farms as well. In some cases, such as for hog finishers in the Philippines, the transaction cost variables mattered greatly for the smallholder sample but had no statistically significant effect on the relative profit efficiency of large farms within the large-farm sample. The interpretation of this result is not that transaction costs did not matter to those large farmers but that all sample members apparently had dealt with them sufficiently, so they did not explain differences in performance among them.

What Is the Outlook for Independent Smallholder Livestock Farming?

Conclusions on competitiveness from the tests of Hypotheses 1 (about relative unit profits) and 2 (about relative profit efficiency) in Chapter 7 indicate that smallholders have a better chance in milk production and in agricultural economies where scaling-up has not progressed to the point it already has in Brazil and Thailand. In some cases, such as for hog farmers in the Philippines, smallholders are actually more competitive in informal local markets than are large-scale farmers, and these informal markets have expanded enough in recent years to allow them to expand their production rapidly. Results for Hypothesis 2 in Chapter 7 also suggest that scaling-up will be different in different places and for different commodities but is likely to continue to occur. Furthermore, as markets gravitate to higher-end concerns of quality and safety, smallholders will increasingly need to be associated with institutions that can supply the technology, inputs, information, and accreditation necessary for them to compete in higher-value markets.

Given the conclusions in Chapter 7 concerning Hypotheses 3 (that smallholders mitigate environmental externalities more) and 4 (that environmental externalities favor large farms), it is clear that promoting action on environmental concerns is not incompatible with promoting small-scale livestock production and that many large-scale producers still need an incentive to clean up their act. It seems plausible that over time, enforcement of environmental regulations will be more similar to enforcement of health regulations, where all producers will be forced to bear the same unit cost regardless of size.

From findings in Chapter 7 pertaining to Hypotheses 5 (that smallholder profits are more sensitive to transaction costs than those of large farmers) and 6 (that contract farmers are more profit-efficient than independents, especially the larger ones), it seems highly likely that monogastric livestock development will continue to see increasing vertical coordination to overcome the high transaction costs faced by all in securing low-cost quality inputs on credit and achieving market recognition for quality outputs. However, contrary to expectations, it is not clear that contact farming will disproportionately favor the interests of small farmers over large ones, and it is quite conceivable that scaling-up will occur in

tandem with and even be promoted through increased vertical coordination within the livestock sector. It seems likely that integrators will have a financial incentive to contract with larger farms if they are free to do so and that the increased profit efficiency of these larger contract farms will eventually capture market share from smallholders. The most likely case for successful smallholder contract farming is likely to be seen where there are few alternatives for processors or integrators to find alternate sources of supply for raw materials.

Implications for Poverty Alleviation and Sustainability Policies

Effects on Poverty via Displacement of Smallholders

The main concern with regard to the forces promoting the scaling-up of livestock production in developing countries is that it might drive small-scale producers out of business altogether in a context in which they would not be able to be absorbed elsewhere in the economy. The issue is not whether such displacement will occur but at what pace and whether it is being sped up artificially by policy distortions, externalities, or structural factors such as transaction costs that have a greater effect on small-scale farms.

The study results in almost all the cases involving pigs and poultry suggest that the smallest independent livestock farms will increasingly have a hard time remaining in business to the extent that their livelihoods depend on that business, although they may remain engaged in part-time activity if the opportunity cost of family labor is low. The results are more encouraging for the next-largest group of smallholders, who remain very small by world standards. The relative profit efficiencies and levels of profit of the "top half of the bottom 40 percent" suggest that they could make a go of livestock production, especially milk production under the current degree of trade protection and especially with some vertical integration in the case of swine. The outlook for small poultry and egg producers is not good.

In the Thai case, it seems that the absolute number of smallholders engaged in market-oriented livestock production has significantly decreased over the past 15 years. Although small-scale livestock farms are mainly found in remote areas today, they continue to account for a significant share of production. As recently as 2003, some 43 percent of broilers produced in Thailand still came from farms keeping fewer than 10,000 birds at a time (Table 2.1), which raises issues on the feasibility of culling strategies to control disease outbreaks such as avian influenza.

Perhaps because the more marginal producers have dropped out in Thailand, the Thai data show that livestock income distribution across farms actually became more equitable between 1993 and 1998 and that the average income of livestock-producing households increased. Possibly those that remained in business were more willing or able to transform as successful market-oriented producers. In any event, for farms engaged in specialized livestock enterprises, the smallest farms, on less than one to two hectares, were the ones that achieved the greatest growth of income.

Anecdotal evidence in Brazil suggests that many of the small-scale swine farms in the temperate south exited the livestock sector after they took out loans to increase the size of their operations at the request of integrators who refused to renew contracts with producers with only a few animals, and then were unable to repay. Integrators were leery of renewing contracts with smallholders because of the increased costs being experienced by the private sector associated with providing technical assistance, delivery of inputs, and pick-up of the final products to many small contract growers. This reinforces the earlier argument that increased contract farming may actually promote scaling-up rather than serving to maintain smallholder farming.

The fact remains that public policy that is targeted to having a widespread effect on poverty through keeping smallholders involved with the growing livestock sector needs to harness the resources of the growing private sector, typically through the provision of an environment that facilitates contract farming. The key for poverty alleviation is to ensure that the form it takes is beneficial to the smallholder growers as well as larger farmers and integrators. Investigating the full costs and benefits of different policies to encourage contracting with smaller-scale farmers is a policy research priority. The incentives for such schemes often come in the form of tax breaks to the integrators, and it will be important to factor in the costs of forgone public revenue when establishing the unit costs of the scheme.

This study did not assess the poverty alleviation possibilities offered by employment opportunities on large-scale livestock farms. However, the existence of positive unit profits where family labor was costed at market wage rates in the Indian and Philippine cases suggest that smallholders may be better off on their own land than working for someone else, or at least better off in working on their own holdings as much as possible. This topic is a priority for future research.

Effects on Resource Degradation through Nutrient Loading

The major sustainability problem associated with the types of growing livestock production studied here concerns poor disposal of the increased amount of manure and dead animals generated in the process. The main environmental problems that policymakers have been concerned with are water pollution, air pollution, and land degradation. Water pollution may occur if nutrients from manure enter the water table because they are either improperly used or disposed of, or it may be associated with improper disposition of dead animals that consequently release nutrients into the groundwater as they decompose. Air pollution may result as the nitrogen in manure is converted to ammonium and also through incineration of dead animals. Land degradation may occur if the carrying capacity of animals is too high, leading to overgrazing in the case of ruminants, or if the application of nutrients over a number of years causes buildup of nitrogen, phosphorus, and salt, resulting in reduced crop yields. Policymakers in the developed countries have begun to address the environmental ills from nutrient loading through stringent enforcement of environmental rules and regulations.

Rules and regulations have also been developed in the different study countries to control potential environmental problems resulting from livestock. Brazil, the Philippines, and Thailand have the most comprehensive sets of rules of the four countries. India has minimal environmental rules regarding livestock. The gist of the matter is that general regulations and institutions for protecting water quality and reducing air pollution have been amended over the years to apply to livestock, particularly to the discharge of wastes from large operations. Furthermore, major livestock product exporters such as Brazil and Thailand have already implemented many changes in rules affecting animal health and have ensured compliance. This was done by state institutions with the full support of large-scale producers, who have a stake in export markets. It seems probable that this trend will continue and will be extended to environmental concerns.

At the same time, all four study countries report problems with environmental enforcement. Thailand's livestock export zone has some of the tightest animal health regulations of any developing country. Yet although a large number of rules and regulations on wastewater management have been developed in Thailand and some incentives are offered to the livestock farms to invest in waste treatment technology,

there has been a lack of enforcement even there. The Thai Department of Livestock and Development (DLD) is expected to be a key enforcer of environmental protection laws, as it currently is of animal health regulations.

While the DLD has the full support of Thailand's large poultry producers and exporters in order to keep foreign markets open through disease control, the consensus might break down on environmental issues. The latter may be less clearly linked to export goals and have the potential for pitting producer interests against those of the general population. Thus, it seems likely that responsible environmental management will require building a broader consensus and using an institutional base that is not subject to conflicts of interest through the DLD's current strong identification with producers.

Similarly, in Brazil, though environmental laws exist, there are problems with the enforcement of these laws due to lack of agreement on the part of farmers with government policies in this area. In certain regions the laws are stricter than in others, and the Brazil study found that farmers were relocating from the highly concentrated livestock production regions in the south, where environmental regulations were increasing, to the center-west, which had fewer environmental regulations and was closer to the inputs to production. Tellingly, of 3,505 agricultural producers covered in a survey done in Brazil in 2001, 76 percent considered the environmental issue one that must be dealt with, yet only 5 percent of the producers approved the current environmental plan of the government.

As the livestock sector has become more industrialized, livestock farms have tended not only to be larger in size, but also to be run more as big businesses. In implementing new environmental technologies, such farm-firms may be able to achieve economies of scale in environmental mitigation, including in unit treatment costs, using lagoons or manure storage sheds. For the most part, they have also tended to be under closer scrutiny by national, state, and local authorities than is the case for smallholders. Regulatory agencies in the Philippines, for example, have in the past cracked down only on large farms. One such agency in the high-density Metro Manila livestock zone is the Laguna Lake Development Authority (LLDA) in Southern Luzon. The resulting trend has been for large-scale operations to move from the periurban areas to less populated areas such as Mindanao. Anecdotal accounts suggest that LLDA is seriously contemplating how to enforce regulations regarding the manure disposal practices of smallholders within its mandate areas.

The study suggests several conclusions for policymakers concerned with environmental sustainability. First, the institutions necessary for the enforcement of environmental standards for smallholders are quite different from those for large farms. It is not reasonable to expect an environmental impact statement from backyard farmers, nor is it easy for a centralized government agency to monitor compliance where tens of thousands of small producers are involved. Instead, management will have to be community-based, with common technical guidelines from a central agency. There also needs to be a means of appeal to a legal authority outside the local community to protect both producers and inhabitants of regions dominated by powerful producers. Second, it is probable that the government will need to find areas closer to inputs to production and farther away from urban areas that are suitable for the scaling-up of the livestock sector that is occurring in their countries and will also need to provide an infrastructure and policy environment that will facilitate the relocation of the industry to these zones.

Effects on Sustainability via Animal Health and Food Safety

Concerns about animal health and food safety are increasingly becoming part of the policy dialogue on the sustainability of livestock farming in developing countries. Recent outbreaks of avian influenza and bovine spongiform encephalopathy and their links to fatal human disease have particularly fueled concerns. Policymakers in these countries are faced with the difficulty of monitoring compliance with disease control measures in a context in which large numbers of smallholders are intermixed with large-scale operations. Large operations, for the most part, have the resources and understanding of the issues necessary to monitor their animals on a continual basis, while smallholders often lack this capacity.

The public sector has typically taken the lead in providing such coverage for smallholders in our study countries, but this has not been enough. Compensation for culled animals is an important aspect of disease control. In the case of the 2004 outbreak of avian flu in Thailand, policymakers choose a stamping-out policy for infected animals that involved the wholesale culling and disposal of animals. The government provided some compensation (less than 50 percent of the market value per bird in most cases). Integrators lost part of the capital they had invested in the birds and feed used, but contractors lost their livelihoods. Independent farmers lost the ability to fully repay loans for DOCs and feeds, and thus possibly future access to credit. Initial control strategies that mandate enclosing poultry in the hot season of a tropical country effectively also eliminate smallholders if enforced, because the evaporative cooling systems used by large farms are prohibitively expensive for smaller ones.

Whether small-scale broiler producers have a chance to remain active participants in the sector, even under contractual arrangements whereby the health of their birds can be monitored by large-scale integrated operations, remains unclear as of this writing. Learning how to better control and share catastrophic animal disease risks under contract farming will be essential to the continuation of this institution involving smallholders in Asia.

At the same time, many governments have been struggling with how to create export opportunities through the development of disease-free export zones for beef and swine products. Control of OIE List A diseases without vaccination is still the standard for exports to the developed countries and increasingly to other countries as well. Control requires effective surveillance. As governments often lack the funds to monitor these regions, lack of public sector capacity has increased the incentive of large-scale export producers to co-opt small-scale producers into a system of disease control that is similar in terms of overhead costs to those practiced on much larger farms.

The Brazilian study reported that producers in certain states that are in the process of being declared FMD-free have organized among themselves to pay into an indemnity fund. The fund will help ensure that a compensation scheme is available to encourage those who may not otherwise do so to report disease outbreaks, because the large producers feel that the government does not have the resources to do so. Although such producer-organized indemnity funds were not described in the Philippine and Thai reports, both of these reports indicated that there was an expansion of large-scale operations to these emerging FMD-free regions. The issue and scope for collective action associating integrators and both large and small farms for animal disease control are policy research priorities.

In all the study countries, there has also been a rising demand for livestock products with specific food safety and quality attributes, probably linked to increased urbanization and income levels. The private sector in the study countries has taken the lead in delivering products with the desired attributes, at least to wealthier consumers patronizing up-market outlets. Large producers in all the study countries have also sought a form of branding through vertical integration with small-scale retail outlets for meat and milk that serve the broader urban populace.

For smallholders to stay involved with this fast-growing segment of the market, they need to meet the evolving food safety standards of these companies and establish credibility for their products with procurement officers. This need typically will require that they integrate into high-value chains through contract farming or other forms of institutional arrangements that have process-based food safety systems in place and can deliver a form of branding.

These types of contractual arrangements ensuring the application of certain quality and food safety standards are increasingly widespread in the vertically coordinating poultry and swine industries of the study countries. Similarly, large-scale dairy producers and dairy cooperatives in the study countries have succeeded in establishing market trust and reputation. Such groups can depend on repeat sales to the same clients who can identify the source of the milk and are willing to pay for something they believe to be of reliable quality and safety. It will be harder for independent smallholders to remain involved over time as markets become more demanding in terms of information about the quality of the product at the time of sale and as market chains become longer and more anonymous.

The existing technologies that ensure food safety may not be scale-neutral, and thus may also be driving scaling-up. For instance, the Brazilian study reported that the chilling and milking technologies required on-farm in recent years to meet the food safety requirements of private processors have made the continued participation by producers of less than 100 liters of milk per day infeasible. Collective action to involve smallholders with the minimum viable scale of such technologies is essential to their continued involvement.

Different approaches to food safety are at play in the different study countries, as noted earlier, principally depending on whether they are large-scale exporters or not. Both Brazil and Thailand use process-based approaches to food safety, such as hazard analysis and critical control point systems, which are similar to systems used in the developed countries and necessary in order to trade with them. The requirements applied and the enforcement of the requirements are largely driven by the private sector. Smallholders will need to find ways to be vertically integrated with certified processing plants if they wish to remain involved in the long term. This will require that they find ways to meet and follow the standards set by processors and integrators.

Failing the emergence of innovations that improve the access of the mass of poor livestock keepers to the rapidly evolving end markets for livestock products in the urban areas, it is hard to see a bright future for smallholder livestock production in fast-growing developing countries. This will require ways to allow smallholders to establish trust and reputation along the value chain. On the other hand, the study results also show that it is unlikely that smallholders will disappear anytime soon, and the animal health examples underline a broader finding that the livestock sector is fundamentally a sector in which all participants sink or swim together.

Finally, while poverty is still mainly a rural phenomenon in developing countries, rising livestock prices can be very harmful to poor urban consumers. There is nothing in the study results that could or should be used to justify public subsidies to livestock producers or to artificially raise the prices of livestock products as an income support measure. Such policies might put off the day of reckoning but would be expensive to implement and ultimately self-defeating if they prevented other actions by livestock producers or public authorities to build a more competitive smallholder link to growing markets.

References

Abdalla, C. W., L. E. Lanyon, and M. C. Hallberg. 1995. What we know about historical trends in firm location decisions and regional shifts: Policy issues for an industrializing animal sector. *American Journal of Agricultural Economics* 77 (5): 1229–1236.

Abdulai, A., and C. L. Delgado. 1999. Determinants of non-farm earnings of farm-based husbands and wives in northern Ghana. *American Journal of Agricultural Economics* 81 (February): 117–130.

Ali, M., and J. C. Flinn. 1989. Profit efficiency among basmati rice producers in Pakistan Punjab. *American Journal of Agricultural Economics* 71 (2): 303–310.

Bates, R. H. 1981. *Markets and states in tropical Africa.* California Series on Social Choice and Political Economy. Berkeley: University of California Press.

Battese, G. E. 1992. Frontier production functions and technical efficiency: A survey of empirical applications in agricultural economics. *Agricultural Economics* 7 (3/4): 185–208.

Battese, G. E., and T. J. Coelli. 1993. A stochastic frontier production function incorporating a model for technical inefficiency effects. Working Papers in Econometrics and Applied Statistics 69, Department of Econometrics, University of New England, Armidale, New South Wales, Australia.

———. 1995. A model for technical inefficiency effects in a stochastic frontier production function for panel data. *Empirical Economics* 20 (2): 325–332.

Battese, G. E., D. S. P. Rao, and C. J. O'Donnell. 2004. A metafrontier production function for estimation of technical efficiencies and technology gaps for firms operating under different technologies. *Journal of Productivity Analysis* 21 (1): 91–103.

Birthal, P. S., and V. K. Taneja. 2006. Livestock sector in India: Opportunities and challenges for smallholders. A paper presented at the international workshop Smallholder Livestock Production in India: Opportunities and Challenges, organized jointly by the Indian Council of Agricultural Research and the International Livestock Research Institute, January 31–February 1, New Delhi, India. Available at http://www.ilri.org/Infoserv/webpub/fulldocs/SmallholderLivestockPro/SmallholderSMPinIndia_Final.pdf.

Camargo Barros, G. S., S. D. Zen, M. R. Piedade Bacchi, S. H. Galvão de Miranda, C. Narrod, and M. Tiongco. 2003. *Policy, technical, and environmental determinants and implications of the scaling-up of swine, broiler, layer and milk production in Brazil.* Final Report of the IFPRI-FAO Livestock Industrialization Project, Phase II, Annex 5. Washington, D.C.: International Food Policy Research Institute. Available at http://www.fao.org/wairdocs/LEAD/x6170e/x6170e00.HTM.

Catelo, M. A. O. 2002. Issues and policy options for addressing environmental human health risks associated with growth in small- and large-scale swine and poultry production in the Philippines. Annex VI, Final Report of IFPRI–FAO Livestock Industrialization Project: Phase I. Washington, D.C.: International Food Policy Research Institute. Available at http://www.fao.org/wairdocs/LEAD/X6115E/x6115e00.HTM.

CEPEA-ESALQ/USP-FAO/LEAD (Center for Advanced Studies on Applied Economics–College of Agriculture "Luiz de Queiroz"/University of São Paulo–Food and Agriculture Organization of the United Nations/Livestock, Environment, and Development Initiative). 2002. CEPEA-ESALQ/

USP-FAO/LEAD Livestock Industrialization field survey dataset 2002. IFPRI-FAO Livestock Industrialization Project: Phase II. Washington, D.C.: International Food Policy Research Institute. CD-ROM.

Coelli, T. J. 1996. A guide to Frontier version 4.1: A computer program for stochastic frontier production and cost function estimation. CEPA Working Papers 7/96. Department of Econometrics, University of New England, Armidale, New South Wales, Australia.

Coelli, T. J., D. S. Prasada Rao, and G. E. Battese. 1998. *An introduction to efficiency and productivity analysis.* Dordrecht: Kluwer Academic.

Costales, A. C., C. Delgado, M. A. O. Catelo, M. Tiongco, A. Chatterjee, A. delos Reyes, and C. Narrod. 2003. Policy, technical, and environmental determinants and implications of the scaling-up of broiler and swine production in The Philippines. Final Report of the IFPRI–FAO Livestock Industrialization Project, Phase II, Annex 1. Washington, D.C.: International Food Policy Research Institute. Available at http://www.fao.org/wairdocs/LEAD/x6170e/x6170e00.HTM.

Costales, A. C., C. Delgado, M. A. O. Catelo, L. Lapar, M. Tiongco, S. Ehui, and A. Z. Bautista. 2007. *Scale and access issues affecting smallholder hog producers in an expanding peri-urban market: Southern Luzon, Philippines.* IFPRI Research Report 151. Washington, D.C.: International Food Policy Research Institute.

De Janvry, A., M. Fafchamps, and E. Sadoulet. 1991. Peasant household behaviour with missing markets: Some paradoxes explained. *Economic Journal* 101 (November): 1400–1417.

Delgado, C. L. 2003. Rising consumption of meat and milk in developing countries has created a new food revolution. *Journal of Nutrition* 133 (11), Supplement 2 on Animal Source Foods: 3907S–3910S.

Delgado, C. L., and N. Minot. 2003. GRP27 proposal—Participation in high-value agricultural markets. Paper presented at the International Food Policy Research Institute, Washington, D.C., June 12.

Delgado C. L., M. Rosegrant, H. Steinfeld, S. Ehui, and C. Courbois. 1999. Livestock to 2020: The next food revolution. Food, Agriculture, and Environment Discussion Paper 28. International Food Policy Research Institute, Washington, D.C.

Deolalikar, A. B., and W. P. M. Vijverberg. 1987. A test of heterogeneity of family and hired labour in Asian agriculture. *Oxford Bulletin of Economics and Statistics* 49 (3): 291–305.

Faasen, H., and H. van Dijk. 1987. Manure as a source of nitrogen and phosphorus in soils. In *Animal manure on grassland and fodder crops: Fertilizer or waste?* ed. H. van der Meer et al. Wageningen, The Netherlands: Martinus Nijhoff.

Fallert, R., M. Weimar, and T. Crawford. 1993. Here's why milk is moving west. *Hoard's Dairy Magazine* 139 (1): 7.

Feder, G. 1985. The relationship between farm size and farm productivity. *Journal of Development Economics* 18 (2–3): 297–313.

Fried, H. O., C. A. Knox Lovell, and S. S. Schmidt, eds. 1993. *The measurement of productive efficiency: Techniques and applications.* New York Oxford University Press. 425 pp.

Gardner B. D., and R. D. Pope. 1978. How is scale and structure determined in agriculture? *American Journal of Agricultural Economics* 60 (2): 295–302.

Government of India. 2000. Annual Report, Department of Animal Husbandry and Dairying, Ministry of Agriculture. New Delhi.

Holloway, G., C. Nicholson, C. Delgado, S. Staal, and S. Ehui. 2000. Agroindustrialization through institutional innovation: Transaction costs, cooperatives and milk-market development in the East-African Highlands. *Agricultural Economics* 23 (3): 279–288.

IIM-IFPRI (Indian Institute of Management–International Food Policy Research Institute). 2003. IIM-IFPRI Livestock Industrialization field survey dataset 2002–03. IFPRI–FAO Livestock Industrialization Project: Phase II. Washington, D.C.: International Food Policy Research Institute. CD-ROM.

Jondrow, J., C. A. Knox Lovell, I. S. Materov, and P. Schmidt. 1982. On the estimation of technical inefficiency in the stochastic frontier production function model. *Journal of Econometrics* 19 (2–3): 233–238.

Kellogg, R. L., C. H. Lander, D. Moffit, and N. Gollehon. 2000. Manure nutrients relative to the capacity of cropland and pastureland to assimilate nutrients: Spatial and temporal trends for the United States. United States Department of Agriculture, NRCS, and ERS, http://www.nrcs.usda .gov/technical/ land/pubs/manntr.html.

Key, N., E. Sadoulet, and A. de Janvry. 2000. Transactions costs and agricultural household supply response. *American Journal of Agricultural Economics* 82 (May): 245–259.

Koester, U. 2004. A revival of large farms in Eastern Europe—How important are institutions? In *Proceedings of the 25th International Conference of Agricultural Economists (IAAE),* 91–102. Durban, South Africa, August 16–22.

Kumbhakar, S. 2001. Estimation of profit functions when profit is not maximum. *American Journal of Agricultural Economics* 83 (1): 1–19.

Kumbhakar, S., and K. Lovell. 2000. *Stochastic frontier analysis.* Cambridge: Cambridge University Press.

Lopez, R. 1986. Structural models of the farm household that allow for interdependent utility and profit maximization decisions. In *Agricultural household models,* ed. I. Singh, L. Squire, and J. Strauss. Baltimore: Johns Hopkins University Press.

Martinez, S. W. 2002. Vertical coordination of marketing systems: Lessons from the poultry, egg, and pork industries. Agricultural Research Report 807, Economic Research Service, United States Department of Agriculture, Washington, D.C.

Mehta, R., R. G. Nambiar, C. Delgado, and S. Subrahmanyam. 2003. Policy, technical, and environmental determinants and implications of the scaling-up of broiler and egg production in India. Final Report of the IFPRI–FAO Livestock Industrialization Project, Phase 2, Annex 2. Washington, D.C.: International Food Policy Research Institute. Available at http://www.fao.org/ wairdocs/LEAD/x6170e/x6170e00.HTM.

Mellor, J. W., and M. S. Mudahar. 1992. Agriculture in economic development: Theories, findings, and challenges in an Asian context. Part Two in *A survey of agricultural economics literature,* Vol. 4: *Agriculture in economic development, 1940s to 1990s,* ed. L. R. Martin, 331–542. Minneapolis: University of Minnesota Press.

Morrison Paul, C., R. Nehring, D. Banker, and A. Somwaru. 2004. Scale economies and efficiency in U.S. agriculture: Are traditional farms history? *Journal of Productivity Analysis* 22 (3): 185–205.

Narrod, C. 1997. Technology transfer in the poultry industry: An examination of supply factors and externalities associated with increased production. Ph.D. dissertation, University of Pennsylvania, Philadelphia.

Nell, A. J., ed. 1998. Livestock and the environment. In *Proceedings of the International Conference on Livestock and the Environment Held in Ede/Wageningen, The Netherlands, June 16–20, 1997.* Wageningen: International Agricultural Centre.

Ollinger M., J. M. MacDonald, and M. Madison. 2005. Technological change and economies of scale in U.S. poultry processing. *American Journal of Agricultural Economics* 87 (1): 116–129.

Outlaw, J., R. Schwart Jr., R. Knutson, A. Pagano, J. Miller, and A. Gray. 1993. Impacts of dairy waste management. Agricultural and Food Policy Center, Policy Working Paper 93-4. College Station: Texas A&M University.

Payne, W., and R. T. Wilson. 1999. *An introduction to animal husbandry in the tropics,* 5th ed. Oxford: Blackwell Science.

Pennsylvania State Cooperative Extension Service. 1993. *The agronomy guide.* University Park: Pennsylvania State University.

Philippines, Bureau of Agricultural Statistics (BAS). 1987. *Backyard livestock and poultry survey: Manual of operations.* Quezon City, Philippines: BAS. Available at http://www.bas.gov.ph/downloads_view.php?id=219.

———. 2007a. *Inventory of native and broiler chicken farms, by region, 1990–2006.* Quezon City, Philippines: BAS. Available at http://www.bas.gov.ph/downloads_view.php?id=219.

———. 2007b. *Inventory of hogs in backyard and commercial farms, by region, 1990–2006.* Quezon City, Philippines: BAS. Available at http://www.bas.gov.ph/downloads_view.php?id=219.

Poapongsakorn, N. 1985. The commercial broiler and swine industries in Thailand. In *Food policy analysis in Thailand,* ed. T. Panayotou, 223–297. Bangkok: Agricultural Development Council.

Poapongsakorn, N., V. NaRanong, C. Delgado, C. Narrod, P. Siriprapanukul, N. Srianant, P. Goolchai, S. Ruangchan, S. Methrsuraruk, T. Jittreekhun, N. Chalermpao, M. Tiongco, and B. Suwankiri. 2003. Policy, technical, and environmental determinants and implications of the scaling-up of swine, broiler, layer and milk production in Thailand. Final Report of IFPRI–FAO Livestock Industrialization Project, Phase II, Annex 4. Washington, D.C.: International Food Policy Research Institute. Available at http://www.fao.org/wairdocs/LEAD/x6170e/x6170e00.HTM.

Reardon, T., and C. P. Timmer. 2005. Transformation of markets for agricultural output in developing countries since 1950: How has thinking changed? Chapter 13 in *Handbook of agricultural economics,* Vol. 3: *Agricultural development: Farmers, farm production and farm markets,* ed. R. E. Evenson, P. Pingali, and T. P. Schultz. Amsterdam: Elsevier.

RISDC-IFPRI (Research and Information System for Developing Countries–International Food Policy Research Institute). 2003. RISDC-IFPRI Livestock Industrialization field survey dataset 2002–3. IFPRI-FAO Livestock Industrialization Project: Phase II. Washington, D.C.: International Food Policy Research Institute. CD-ROM.

Rosenzweig, M. R. 1988. Labor markets in low-income countries. Chapter 15 in *Handbook of Development Economics,* Vol. 1, ed. H. Chenery and T. N. Srinivasan, 714–762. Amsterdam: Elsevier.

Schwart, R., J. Holt, and J. Outlaw. 1995. Economic factors driving regional shifts in livestock production: Opportunities for policy innovation. In *Animal waste and the land-water interface,* ed. K. Steele. Boca Raton, Fla., U.S.A.: Lewis Publications.

Sere, C., and H. Steinfeld. 1996. World livestock production systems: Current status, issues and trends. FAO Animal Production and Health Paper 127. Paper prepared in collaboration with Jan Groenewold. Food and Agriculture Organization of the United Nations, Rome.

Sharma, V. P., S. Staal, C. Delgado, and R. V. Singh. 2003. Policy, technical, and environmental determinants and implications of the scaling-up of milk production in India. Research Report of the IFPRI–FAO Livestock Industrialization Project, Phase 2, Annex 3. Washington, D.C.: International Food Policy Research Institute. Available at http://www.fao.org/wairdocs/LEAD/x6170e/x6170e00.HTM.

Sims, T. 1995. Characteristics of animal wastes and waste-amended soils: An overview of the agricultural and environmental issues. In *Animal wastes and the land-water interface,* ed. K. Steele. Boca Raton, Fla.: Lewis Publications.

Sims, T., and D. Wolf. 1994. Poultry waste management: Agricultural and environmental issues. In *Advances in agronomy,* vol. 52, ed. D. Sparks. Orlando, Fla., U.S.A.: Academic Press.

Singh, I., L. Squire, and J. Strauss, eds. 1986. *Agricultural household models: Extensions, applications, and policy.* Baltimore: Published for the World Bank by Johns Hopkins University Press.

Staal, S., C. Delgado, and C. Nicholson. 1997. Smallholder dairying under transactions costs in East Africa. *World Development* 25 (5): 779–794.

StataCorp. 2006. *Stata statistical software: Release 9.2.* College Station, Tex., U.S.A.: Stata Corporation.

Steinfeld, H., C. de Haan, and H. Blackburn. 1997. Livestock-environment interactions: Issues and options. Report of a study sponsored by the Commission of the European Communities, the World Bank, and the governments of Denmark, France, Germany, Netherlands, United Kingdom, and United States. The study was coordinated by the Food and Agriculture Organization of the United Nations, the United States Agency for International Development, and the World Bank. Available at http://lead.virtualcenter.org/en/dec/toolbox/FAO/Summary/index.htm.

Stoyke, C., and H. Waibel. 1994. De-intensification strategies of cereal farms in a contribution toward more sustainable farming systems. In *Rural and farming systems analysis: European perspectives,* ed. J. B. Dent and M. McGregor. Wallingford, U.K.: CABI.

TDRI-FAO/LEAD (Thailand Development Research Institute–Food and Agriculture Organization of the United Nations/Livestock, Environment, and Development Initiative). 2002. TDRI-FAO/LEAD Livestock Industrialization field survey dataset 2002. IFPRI-FAO Livestock Industrialization Project: Phase II. Washington, D.C.: International Food Policy Research Institute. CD-ROM.

Thailand, National Statistical Office. 1988. *Intercensal survey of agriculture.* Bangkok, Thailand: National Statistical Office.

———. 1993. *Agricultural census.* Bangkok, Thailand: National Statistical Office.

———. 1998. *Intercensal survey of agriculture.* Bangkok, Thailand: National Statistical Office.

———. 2003. *Agricultural census.* Bangkok, Thailand: National Statistical Office.

Tokle, J. G., and W. E. Huffman. 1991. Local economic conditions and wage labor decisions of farm and rural non-farm couples. *American Journal of Agricultural Economics* 73 (3): 652–670.

UPLB-IFPRI (University of the Philippines Los Baños–International Food Policy Research Institute). 2003. UPLB-IFPRI Livestock Industrialization field survey dataset 2002–3. IFPRI-FAO Livestock Industrialization Project: Phase II. Washington, D.C.: International Food Policy Research Institute. CD-ROM.

Westenbarger, D., and D. Letson. 1995. Livestock and poultry waste-control costs. *Choices* 10 (2): 27–30.

Williams, P. 1995. Animal production and European pollution problems. *Animal Feed Science and Technology* 53 (2): 135–144.

Williamson, O. E. 1989. Transaction cost economics. In *Handbook of industrial organization,* Vol. 1, ed. R. Schmalensee and R. D. Willig. Amsterdam: Elsevier.

World Bank. 2002. *World development report: Sustainable development in a dynamic world.* Washington, D.C.